De-signing Design

TEXTURES: Philosophy / Literature / Culture
Series Editor: John William Phillips, National University of Singapore

This series seeks to publish the most exciting in-depth research in the areas of Philosophy, Literature, and Culture today. TEXTURES has been established to include not only contemporary interdisciplinary studies in philosophy, literature, film, media, and the arts, but also literary, aesthetic, and cultural theory. It addresses questions of cultural meaning and cultural difference, aesthetic experience and cultural studies while focusing on new directions in philosophical/ literary/ art/ musical/ film/ cultural theory.

This series sets a new standard for quality books in the interrelations between philosophy, literature, the arts, and culture and for identifying some of the most important and pressing contemporary issues in these inter-cultural and cross-disciplinary areas. Volumes are to emphasize the intersections between disciplinary practices and the ways in which these differences in practices can be thematized and articulated theoretically and philosophically. They should provide a focused contribution to varying aspects of a contemporary or thematic topic.

Titles in the Series

Dramas of Culture: Theory, History, Performance, edited by Wayne Jeffrey Froman and John Burt Foster, Jr.

Intermedialities: Philosophy, Arts, Politics, edited by Henk Oosterling and Ewa Plonowska Ziarek

De-signing Design: Cartographies of Theory and Practice, edited by Elizabeth Grierson, Harriet Edquist, and Hélène Frichot

De-signing Design

Cartographies of Theory and Practice

Edited by
Elizabeth Grierson,
Harriet Edquist, and Hélène Frichot

Editor-in-Chief Hugh J. Silverman

LEXINGTON BOOKS
Lanham • Boulder • New York • London

Published by Lexington Books
An imprint of The Rowman & Littlefield Publishing Group, Inc.
4501 Forbes Boulevard, Suite 200, Lanham, Maryland 20706
www.rowman.com

Unit A, Whitacre Mews, 26-34 Stannary Street, London SE11 4AB

British Library Cataloguing in Publication Information Available

Library of Congress Cataloging-in-Publication Data

De-signing design : cartographies of theory and practice / Edited by Elizabeth Grierson, Harriet Edquist, and Hélène Frichot.
pages cm. -- (Textures: philosophy / literature / culture)
Includes bibliographical references and index.
ISBN 978-0-7391-7912-3 (cloth : alk. paper) -- ISBN 978-1-4985-1035-6 (pbk. : alk. paper) -- ISBN 978-0-7391-7913-0 (electronic)
1. Art and design. I. Grierson, Elizabeth, editor. II. Edquist, Harriet, editor. III. Frichot, Hélène, editor.
NK1505.D28 2015
745.4--dc23
2015026130

∞™ The paper used in this publication meets the minimum requirements of American National Standard for Information Sciences Permanence of Paper for Printed Library Materials, ANSI/NISO Z39.48-1992.

Printed in the United States of America

Contents

Acknowledgments

Elizabeth Grierson, Harriet Edquist, and Hélène Frichot acknowledge the Design Research Institute of Royal Melbourne Institute of Technology (RMIT) University, for fostering the trans-disciplinary approach adopted in this book and for sponsoring this project. The editors acknowledge the late Professor Hugh Silverman and organizers of *Global Arts/Local Knowledge*, the International Association of Philosophy and Literature (IAPL) 2008 conference at RMIT, for their invitation to present the Design Research Institute plenary panel on the final day of the conference. From that event, the editors extended the research and invited a range of scholars to contribute to the book's designated themes: De-signing Design; Geo-Placed Knowledges and Design; and Mapping a New Biological Paradigm for Design. Thanks to Dr. Neal Haslem, Dr. Tintin Wulia, Rupa Ramanathan, and Virginia Grierson for manuscript assistance; and grateful thanks to all the writers for their insightful and scholarly contributions to *De-signing Design: Cartographies of Theory and Practice*.

Abbreviations

A: Sebald, W. G. *Austerlitz*. New York: The Modern Library, 2001.

AA: Rendell, Jane. *Art and Architecture: A Place Between*. London: I. B. Tauris, 2006.

AAH: Ingraham, Catherine. *Architecture, Animal, Human: The Asymmetrical Condition*. London: Routledge, 2006.

AB: Haweis, Mary Eliza. *The Art of Beauty*. Boston: Adamant Media, 1878/2005.

ACT: Beilby, Raymond, and Cecil Hadgraft. *Ada Cambridge, Tasma and Rosa Praed*. Melbourne: Oxford University Press, 1979.

AE: Weinstock, Michael. *The Architecture of Emergence: The Evolution of Form in Nature and Civilisation*. London: Wiley, 2010.

AH: Chase, Athol. "Thomson Time." *Aboriginal History* 3 (1979): 109–10.

AJA: Hamby, Louise. "A Question of Time: Ten Canoes." *The Australian Journal of Anthropology* 18, no. 1 (2007): 123–26.

AM: Sontag, Susan. *Aids and Its Metaphors*. London: Penguin, 1988.

AP: Pinney, Christopher. "The Parallel Histories of Anthropology and Photography." In *Anthropology and Photography 1860–1920*. Edited by Elizabeth Edwards. New Haven, CT: Yale University Press, 1992.

ASE: Lenoir, Timothy, and Casey Alt. "Flow, Process, Fold." In *Architecture and the Sciences: Exchanging Metaphors*. Edited by Antoine Picon and Alessandra Ponte, 313–52. New York: Princeton University Press, 2003.

AWT:: Agrest, Diana I. "The Misfortunes of Theory." In *Architecture from Without: Theoretical Framings for a Critical Practice*. Cambridge, MA: MIT Press, 1991.

BD: Negroponte, Nicholas. *Being Digital*. New York: Vintage, 1995.

BS: Salomon, Mandy. "Business in Second Life." Paper presented as Smart Internet Technology CRC Research Paper. Last modified December 22, 2010. Accessed June 18, 2008. http://www.smartservicescrc.com.au/Outcomes.html.

BT: Taylor, William. "Building on Transience: Tolerance and the Subjective Dimension of Technology." Paper presented at *Techniques and Technologies, Transfers and Transformation, Association of Australasian Schools of Architecture (AASA) International Education Conference*. Sydney: University of Technology Sydney, 2007.

C: Castner, Henry W. "Might There Be a Suzuki Method in Cartographic Education?" *Cartographica* 18, no. 1 (1981): 59–67.

CAR: Grierson, Elizabeth. "Ways of Deconstructing." In *Creative Arts Research: Narratives of Methodologies and Practices*, 149–63. Rotterdam: Sense Publishers, 2009.

CBF: Lynn, Greg, and Ingeborg M. Rocker. "Calculus Based Form: An Interview with Greg Lynn." *Programming Cultures, AD* 76, no. 4 (July/August, 2006).

CCC: Florida, Richard. *Cities and the CreativeClass*. London/New York: Routledge, 2005.

CCD: Rystedt, B. "Compact Disks for Distribution of Maps and Other Geographic Information." *Proceedings of 13th ICC Conference, Morelia, Mexico*, International Cartographic Association (ICA) IV (1987): 479–84.

CEA: Guattari, Félix. *Chaosmosis: An Ethico-Aesthetic Paradigm*. Translated by Paul Bains and Julian Pefanis. Sydney: Power Publications, 1995.

CJ: Kant, Immanuel. *The Critique of Judgement*. Translated by James Creed Meredith. Oxford: Clarendon Press, 1952.

CL: Barthes, Roland. *Camera Lucida: Reflections on Photography*. Translated by Richard Howard. London: Vintage, 2000.

CLA: Peters, K. "The Contemporary Landscape in Art." Paper presented at *AAG 2007 Annual Meeting*, San Francisco, April 17-21, 2007.

CLR: Barthes, Roland. *Camera Lucida: Reflections on Photography*. Translated by Richard Howard. New York: Hill and Wang, 1981.

CPA: Lynn, Greg. "Greg Lynn: Embryological Houses." *Contemporary Processes in Architecture AD* 70, no. 3 (2000): 26–35.

CWC: Agamben, Giorgio. "Beyond Human Rights." In *Cities without Citizens*. Edited by Eduardo Cadava, Aaron Levy, and Giorgio Agamben, 3–19. Philadelphia with the Rosenbach Museum and Library: Slought Books, Theory Series no. 1, 2003.

DLM: Sharp, Darren. "Digital Lifestyles Monitor." Paper presented as Smart Internet Technology CRC Research Paper. Last modified December 22, 2010. http://www.smartservicescrc.com.au/Outcomes.html.

DTA: Thomson, Donald. *Donald Thomson in Arnhem Land*. Melbourne: The Miegunyah Press, 2006.

DTM: Allen, Lindy. "Photographer of Brilliance." In *Donald Thomson: The Man and the Scholar*. Edited by Bruce Rigsby and Nicolas Peterson, 45–62. Canberra: The Academy of the Social Sciences in Australia, 2005.

EFD: Jablonka, Eve, and Marion J. Lamb. *Evolution in Four Dimensions*. Cambridge, MA: MIT Press, 2005.

EK: Foray, Dominique. *The Economics of Knowledge*. Cambridge, MA: MIT Press, 2004.

EMD: Hensel, Michael. Achim Menges and Michael Weinstock, eds. *Emergence: Morphogenetic Design Strategies AD* 74, no. 3 (May/June, 2004).

EVE: Greenhalgh, Paul. *Ephemeral Vistas: The Expositions Universelles, Great Exhibitions and World Fairs, 1851–1939*. Manchester: Manchester University Press, 1988.

FES: Burchell, Graham, Colin Gordon, and Peter Miller, eds. *The Foucault Effect: Studies in Governmentality*. Chicago: University of Chicago Press, 1991.

FF: Grierson, Elizabeth. "Utopia for All Seasons: Futuring Art—Futuring Education." In *The Chicago Project: Fold-Out Futures II*. Edited by Irene Barberis, 7–14. Melbourne: RMIT Publishing, 2008.

G: Derrida, Jacques. *Of Grammatology*. Translated by Gayatri Chakravorty Spivak. Baltimore and London: The John Hopkins University Press, 1998.

GLL: Assmann, Aleida. "On the (In)compatibility of Guilt and Suffering in German Memory." *German Life and Letters* 59, no. 2 (April 2006): 187–200.

GR: Baer, Ulrich. "To Give Memory a Place: Holocaust Photography and the Landscape Tradition." *Representations* 69, Special Issue: Grounds for Remembering (Winter 2000): 38–62.

HC: Arendt, Hannah. *The Human Condition*. Chicago: University of Chicago Press, 1958.

HDM: Martin, Reinhold. "Critical of What?: Toward a Utopian Realism." *Harvard Design Magazine* 22 (Spring/Summer 2005): 2.

HH: Heidegger, Martin. *Hölderlin's Hymn* "The Ister." Translated by William McNeill and Julia Davis. Bloomington/ Indianapolis: Indiana University Press, 1996.

HS: Foucault, Michel. *The History of Sexuality*. Translated by Robert Hurley. Harmondsworth, Middlesex, England: Penguin, 1990 rpt.

I: Arendt, Hannah, ed. *Illuminations*. Translated by Harry Zohn. London: Fontana, 1973.

IAM: Allen, Lindy. "Tons and Tons of Valuable Material." In *The Makers and Making of Indigenous Australian Museum Collections*. Edited by

Nicolas Peterson, Lindy Allen, Louise Hamby, and Museum Victoria, 387–419. Melbourne: Melbourne University Press, 2008.

IC: Calvino, Italo. *Invisible Cities*. London: Picador, 1979.

IMT: Barthes, Roland. "Introduction to the Structural Analysis of Narratives." In *Image-Music-Text*. Translated by Stephen Heath, 79–124. Glasgow: Fontana, 1977.

L: Carpo, Mario. "Tempest in a Teapot." *Log* 6 (2005): 99–106.

LNM: Manovich, Lev. *The Language of New Media*. Cambridge, MA: Technology Press, 2001.

MA: Frascari, Marco. *Monsters of Architecture: Anthropomorphism in Architectural Theory*. Lanham, MD: Rowman & Littlefield Publishers, 1991.

MB: Goldschmidt, Richard. *The Material Basis of Evolution*. New Haven, CT: Yale University Press, 1940.

MFL: Foucault, Michel. *Language, Counter-memory, Practice: Selected Essays and Interviews*. Edited by Donald F. Bouchard. Ithaca, New York: Cornell University Press, 1996.

MI: Sparke, Penny. *The Modern Interior*. London: Reaktion Books, 2008.

MMD: Stewart, Susan. "Prologue: From the Museum of Touch." In *Material Memories, Design and Evocation*. Edited by Marius Kwint, Christopher Breward, and Jeremy Aynesley. London: Berg, 1999.

MP: Derrida, Jacques. "Différance." In *Margins of Philosophy*. Translated by Alan Bass, 1–27. Sussex: The Harvester Press, 1982.

MSC: Peers, Laura, and Alison K. Brown, eds. *Museum and Source Communities: A Routledge Reader*. London: Routledge, 2003.

MTC: Lakoff, George, and Mark Turner. *More Than Cool Reason*. Chicago/London: University of Chicago Press, 1989.

MWD: Hardt, Michael, and Antonio Negri. *Multitude: War and Democracy in the Age of Empire*. New York: Penguin, 2004.

NAP: Lemm, Vanessa. *Nietzsche's Animal Philosophy: Culture Politics and the Animality of the Human Being*. New York: Fordham University Press, 2009.

NIT: Bal, Mieke. *Narratology: Introduction to the Theory of Narrative*. Toronto: University of Toronto Press, 1997.

NMD: Everett, Anna, and John Caldwell, eds. *New Media, Theories and Practices of Digitextuality*. New York: Routledge, 2003.

NOX: Spuybroek, Lars. *NOX: Machining Architecture*. London: Thames & Hudson, 2004.

NSE: Braidotti, Rosi. *Nomadic Subjects: Embodiment and Sexual Difference in Contemporary Feminist Theory*. New York: Columbia University Press, 1994.

OTN: Derrida, Jacques. *On The Name*. Edited by Thomas Dutoit. Translated by David Wood, John P. Leavy, and Ian McLeod. Stanford: Stanford University Press, 1995.

OD: Willis, Anne M. "Ontological Designing." *Design Philosophy Papers* 2 (2006): 1–11.

OS: Darwin, Charles. *On the Origin of Species By Means of Natural Selection, or the Preservation of Favoured Races in the Struggle for Life*. Cambridge, MA: Harvard University Press, 1966.

OW: Eco, Umberto. *The Open Work*. Translated by Anna Cancogni. Cambridge, MA: Harvard University Press, 1989.

PE: Malpas, J. E. *Place and Experience*. Cambridge: Cambridge University Press, 1999.

PEL: Certeau, Michel de. *The Practice of Everyday Life*. Berkeley: University of California Press, 1984.

PFA: Zaera-Polo, Alejandro, and Farshid Moussavi (Foreign Office Architects). *Phylogenesis: FOA's Ark*. Barcelona: Actar, 2003.

PI: Le Dœuff, Michèle. *The Philosophical Imaginary*. Translated by Colin Gordon. London: The Athlone Press, 1989.

PM: Friedman, Yona. *Programmes and Manifestoes on Twentieth-Century Architecture*. Edited by Ulrich Conrads. London: Lund Humphries, 1970.

PMB: Hayes, Gary. "What Have You Got against My Bot?" Personalize Media blog. Last modified February 18, 2008. http://www.personalizemedia.com/index.php/2008/02/18/377/.

POH: Pinney, Christopher. "Introduction: How the Other Half . . ." In *Photography's Other Histories*. Edited by Christopher Pinney and Nicolas Peterson. Durham and London: Duke University Press, 2003.

POK: Grierson, Elizabeth. "The Politics of Knowledge: A Poststructuralist Approach to Visual Arts Education in Tertiary Sites." PhD dissertation, The University of Auckland, New Zealand, 2000.

POS: Hegel, G. W. F. *Phenomenology of Spirit*. Translated by Arnold V. Miller. Oxford: Oxford University Press, 1977.

PP: Colomina, Beatriz. *Privacy and Publicity: Modern Architecture as Mass Media*. Cambridge, MA: MIT Press, 1994.

PPU: Huyssen, Andreas. *Present Pasts: Urban Palimpsests and the Politics of Memory*. Stanford: Stanford University Press, 2003.

PS: Lefèbvre, Henri. *The Production of Space*. Translated by Donald Nicholson-Smith. Oxford: Blackwell, 1991.

PT: Hirsch, Marianne. "Past Lives: Postmemories in Exile." *Poetics Today* 17, no. 4, *Creativity and Exile: European/American Perspectives* II (Winter 1996): 659–86.

PVL: Morris, William. *Prose, Verse, Lectures and Essays*. Edited by G. D. H. Cole. London: Nonesuch Press, 1946.

QCT: Heidegger, Martin. *The Question Concerning Technology and Other Essays*. Translated by William Lovitt. New York: Harper & Row, 1977.

REA: Benjamin, Walter. "On Language as Such and on the Language of Man." In *Reflections: Essays, Aphorisms, Autobiographical Writings*. Edited by Peter Demetz. New York: Schocken Books, 1986.

RFM: Davison, Graeme. *The Rise and Fall of Marvellous Melbourne*. Melbourne: Melbourne University Press, 1981.

ROP: Mason, J. *Researching Your Own Practice: The Discipline of Noticing*. London: RoutledgeFalmer, 2002.

RR: Eco, Umberto. *The Role of the Reader: Explorations in the Semiotics of Texts*. Bloomington: Indiana University Press, 1979.

S: Baudrillard, Jean. *Simulations*. New York: Semiotext(e), 1983.

SAH: Orr, Kirsten. "A Force for Urbanism and National Identity: Nineteenth-Century Australian International Exhibitions and Their Domestic Exhibits." *Proceedings of XIXth Conference of the Society of Architectural Historians Australia and New Zealand (SAHANZ)*, Brisbane, October 4-7, 2002.

SC: Lozano-Hemmer, Raphael. "Interview by José Luis Barrios." *Subsculptures*. Translated by Rebecca MacSween. Last modified June 10, 2014, http://www.lozano-hemmer.com/texts/bibliography/articles_interviews_essays/Subsculptures_2005jlb.pdf.

SHB: Blanchot, Maurice. *The Station Hill Blanchot Reader, Fiction & Literary Essays*. Edited by George Quasha. Translated by Lydia Davis, Paul Auster, and Robert Lamberton. Barrytown: Station Hill, 1999.

SL: Blanchot, Maurice. *The Space of Literature*. Translated by Ann Smock. Lincoln & London: University of Nebraska Press, 1993.

SLO: Rymaszewski, Michael, James Au Wagner, Mark Wallace, Catherine Winters, Cory Ondrejka, and Benjamin Batstone-Cunningham. *Second Life: The Official Guide*. New Jersey: Wiley, 2007.

SM: Lévi-Strauss, Claude. *The Savage Mind*. Chicago: University of Chicago, 1968.

SNB: Blanchot, Maurice. *The Step Not Beyond*. Translated by Lycette Nelson. Albany: SUNY Press, 1992.

SR: Chen, Qiulin. "Salvaged from Ruins." In the proceedings of *The 6th Asia Pacific Triennial of Contemporary Art*, 75–76. Brisbane: Queensland Art Gallery, 2009.

SSM: Deger, Jennifer. *Shimmering Screens: Making Media in an Aboriginal Community*. Minneapolis: University of Minnesota Press, 2006.

STP: Foucault, Michel. *Security, Territory, Population: Lectures at the Collège de France, 1977-1978*. Edited by Michel Senellart. Translated by Graham Burchell. Hampshire and New York: Palgrave Macmillan, 2007.

SWA: Charvet, P. E. trans. *Charles Baudelaire: Selected Writings on Art and Literature*, New York: Viking, 1972, 395–422, in "The Painter of Modern Life. Charles Baudelaire (1863)," Modernism och postmodernism. Last modified December 22, 2010, http://www.idehist.uu.se/distans/ilmh/pm/baudelaire-painte.htm.

TCK: Ten Canoes. "Press Kit." Palace Films Australia, 2006. Last modified July 1, 2010, http://www.tencanoes.com.au/tencanoes/info.htm.

TCS: Davis, Therese. "Remembering Our Ancestors: Cross-cultural Collaborations and the Mediation of Aboriginal Culture and History in Ten Canoes." *Studies in Australasian Cinema* 1 (2007): 5–14.

TIC: Blanchot, Maurice. *The Infinite Conversation*. Translated by Susan Hanson. Minneapolis: University of Minnesota Press, 1993.

TPC: Deleuze, Gilles, and Félix Guattari. *A Thousand Plateaus: Capitalism and Schizophrenia*. Translated by Brian Massumi. Minneapolis: University of Minnesota Press, 1987.

TPE: Casey, Edward. S. "Body, Self, and Landscape: A Geo-Philosophical Inquiry into the Place-World." In *Textures of Place: Exploring Humanist Geography*. Edited by Paul C. Adams, Steven D. Hoelscher, and Karen E. Till, 403–25. Minnesota: University of Minnesota Press, 2001.

TTM: Hensel, Michael, Achim Menges, and Michael Weinstock, eds. *Techniques and Technologies in Morphogenetic Design AD* 76, no. 2 (March/April 2006).

TYN: DeLanda, Manuel. *A Thousand Years of Nonlinear History*. New York: Zone Books/ Swerve Editions, 1997.

UC: Blanchot, Maurice. *The Unavowable Community*. Translated by Pierre Joris. Barrytown: Station Hill Press, 1988.

VC: Briggs, Asa. *Victorian Cities*. Harmondsworth, Middlesex, England: Penguin, 1985.

VR: Canguilhem, Georges. *The Vital Rationalist*. Edited by François Delaporte. New York: Zone Books, 2000.

VV: Hensel, Michael, and Achim Menges. *Versatility and Vicissitude: Performance in Morpho-Ecological Design AD* 78, no. 2 (March/April 2008).

W: Norris, Christopher, and Andrew Benjamin. *What is Deconstruction?* London: Academy Editions, 1996.

WB: Forty, Adrian. *Words and Buildings: A Vocabulary of Modern Architecture*. New York: Thames & Hudson, 2000.

WD: Derrida, Jacques. *Writing and Difference*. Translated by Alan Bass. Chicago: University of Chicago Press, 1978.

WF: Cambridge, Ada. *A Woman's Friendship*. Edited by Elizabeth Morrison. Kensington: New South Wales University Press, 1988.

WN: Bortoft, Henri. *The Wholeness of Nature: Goethe's Way of Science*. New York: Floris Books, 1996.

WP: Deleuze, Gilles, and Félix Guattari. *What is Philosophy?* Translated by Graham Burchell and Hugh Tomlinson. London: Verso, 1994.

WVS: Eisenman, Peter. *Written into the Void: Selected Writing 1990–2004*. New Haven, CT: Yale University Press, 2007.

YJC: Hirsch, Marianne. "Surviving Images: Holocaust Photographs and the Work of Post-Memory." *Yale Journal of Criticism* 14, no. 1 (2001): 5–37.

ZAZ: Kasulis, T. P. *Zen Action/Zen Person*. Honolulu: University Press of Hawaii, 1981.

ZI: Crary, Jonathan, and Sanford Kwinter, eds. *Zone 6: Incorporations*. New York: Zone Books, 1992.

Introduction

De-signing Design: Cartographies of Theory and Practice

Elizabeth Grierson, Harriet Edquist, and
Hélène Frichot

De-signing Design: Cartographies of Theory and Practice explores the terrain oscillating between thinking and doing or between theory and practice in design and art as discourses, practices, and disciplines: a terrain that can be both productive and perilous. The collection seeks a critical methodology as a way of prising open not only the practices of design and art, but also the discourses of spatiality and urbanization, and the scene of writing itself. Through a discursive set of engagements, the collection is seeking a more critical understanding of spatiality, difference, and identity in the expanded field of designing, place-making, and being. In so doing, it envisages the activations of design as a series of cartographic devices, a set of moves in the interstices between theory and practice. Thus, the collection is not setting up theory and practice as binary poles nor is there an opposition between art and design as practices. The primary interest finds its location in the relations, deferrals, and dispersals that are activating a way of thinking and acting in the world. Thus, the writing considers designing as a critical way of practising, and de-signing as an artful way of signing and thinking about design in the practices of place-making and spatial relationships of urban living.

With these guiding principles in mind, the book draws together three trajectories of investigation that treat contemporary design, research, thinking, and practice in different ways. The writing positions *design* as a discourse that relays across architecture, art, aesthetics, ethics, literature, political governance, urban planning, cartography, and biological sciences. Through trans-disciplinary relationships, design reveals a critical way of thinking as it negotiates personal and public environments and concerns. The collection has three guiding themes that allow the parts to disclose their performative relations to the whole. Firstly, there is *De-signing Design*, then *Geo-Placed Knowledges and Design*, and finally, *Mapping a New Biological Paradigm for Design.* These engagements effectively map distinct navigational directions for reconsiderations of design practice and thinking. To reflect the three key thematic areas the book is organised into three sections, each featuring an introductory essay by one of the coeditors, Elizabeth Grierson, Harriet Edquist, and Hélène Frichot. The three introductions aim to situate the themes more generally in the field of design, and between practice and theory, while also focusing the reader's attention on particular concerns and strategies addressed by the contributors in each part.

It was at the International Association for Philosophy and Literature (IAPL) 2008 conference, *Global Arts/Local Knowledge*, convened in Melbourne, Australia, that the ideas informing these essays first found a place for discussion. The closing roundtable, plenary session, sponsored by the Design Research Institute, RMIT University[1] took the form of a panel of speakers addressing questions of *Design Research Intervening at the Scale of the Local and Global.* Questions and challenges raised by the notion of *de-signing design* as a deconstructive approach to dominant discourses of design led to further conversations between us, that is to say, between Grierson, Edquist and Frichot, as we began the work of collecting and editing the chapters gathered here.

The title, *De-signing Design* presents a challenge to the design disciplines and art, as an aesthetic activation of design thinking and practice, to situate their various endeavours in a terrain of unstable signs for navigating new directions. The task of design and art is to construct novel signs toward thinking and doing that allows us both to continue creating a world, and to see our existing situation in a new light. De-signing design as a mode of active and ongoing engagement in a world of ideas, things and social spaces is suggestive of a process that must always find its way from the midst of things. It is a way of venturing and finding pathways through problems that often arise from contingent encounters, fragmentary conversations and partial relations. De-signing challenges habitual approaches to design, and demands that thinking and doing operate concurrently, and that a relay between these approaches or what can be called a "thinking-doing" is sustained on our travels into the unexpected territories of design and art, practice and theory.

The hyphen in *de-signing* draws from French philosopher, Jacques Derrida, as a point of provocation. Design, without the hyphen, assumes the stability of intentional acts, that design has an object or else a subject, an origin and an end in mind, and that it can communicate its purpose in a reasonably transparent way. Derrida has shown, however, that there is more at play and certainly more noise in the work of signs and the process of signification than that. "De-signed artefacts" communicate signs that do not represent the intention of the designer in a straightforward way, but in their very circulation, they operate through divergence in a discursive process between maker and user. In the discursive process delay, deferral and displacement are in operation. Derrida coined the term, *différance*[2] as a creative force. *Différance* traces pathways of signs that are mapped by way of a sometimes "blind tactics," following an "empirical wandering" that does not guess, in advance, what encounters will be forthcoming (*MP*).

Elizabeth Grierson, in her opening chapter "De-signing the City: Interventions through Art," frames the process of de-signing as a deconstructive approach to interrogating assumptions in the rhetoric of innovation and knowledge transfer as she considers what de-signing might mean for the creative disciplines in an urban setting. Other writers gathered here draw on the influence of poststructuralist philosophers including Jacques Derrida, Michel Foucault, Gilles Deleuze and Félix Guattari, Maurice Blanchot, Jean Baudrillard, Umberto Eco, Michèle Le Dœuff, and others. These philosophical thinkers unsettle the stability of structures of meaning and signification questioning, persistently as well as creatively, the status quo. The practical concerns that drive the art and design disciplines still find the conceptual frameworks offered by such thinkers valuable, even in the face of serial crises, including the pervasiveness of new technologies that question the relevance of theory.[3] What these thinkers offer is some guidance in the mapping and critiquing, as well as constructing of provisional signs of design.

The book also addresses mapping, or more specifically cartography, as a specific methodology. William Cartwright in his chapter, "Representing the City: Complementing Science and Technology with Art," asks how can the new technologies of mapping, which now tend to be located in the disciplines of mathematics and science, information and communication technologies, be re-mapped, or perhaps de-signed, by way of literature, design and art? As Harriet Edquist shows in "Mapping Modernity in 'Marvellous Melbourne': Ada Cambridge's *A Woman's Friendship*," the mapping of different locales depicted in a novel allows literature to be re-engaged as another approach to place-making that emerges in the narration of historical characters and settings. Literature and mapping come together as a way of negotiating place and time, bringing arts and sciences together through new ways of de-signing and navigating the signs of place.

A further preoccupation is located in the way design engages with emerging technologies, coupled with a continuing reliance on old technologies—that is to say, a preoccupation with the interleaving of technologies, depending on the situation of their deployment: with new technologies not necessarily superseding old technologies. For example, Linda Daley's chapter, "Embodied Encounters: The Photographic Seeds of Rolf de Heer and Peter Djigirr's *Ten Canoes*" places the ancient craft of carving canoes by the Yolgnu peoples of Arnhem Land, Australia, alongside the emerging art of exploration photography, which is offered as a further narrative framework through new technologies associated with the cinema. Scott McQuire's chapter, "Toward De-signing: Narratives, Networks and the Open Work" is another that rethinks the issue of narrative in relation to the new, in this case new media. Many of the chapters demonstrate how different technologies as well as cultures and ways of seeing may come together in creative relationships. Often inspired by hybrids of new and old technologies, design may manifest the inauguration of the monstrous, like a de-signed sign operating between horror and wonder, as in Chris Smith's chapter, "Hopeful: Biology, Architectural Design and Philosophy."

In the first part of *De-signing Design*, a focus on the deconstructive activity of de-signing design suggests that we make sense of things anew and this generally frames the ethos of the book. The second part focusing on framing geo-placed knowledges, ventures new approaches to the mapping of existing terrains, real and imagined, as well as the mapping of worlds yet to come. The third section explores and questions a new biological paradigm for design in light of pressing ethical and aesthetic issues. These three threads of concern guide us into unknown territories, in which the making of new maps may occur in the light of a perpetual unfolding of new experiences. *De-signing Design* offers different approaches to an experimental and experiential "thinking-doing" that promises to open up further research possibilities in the fields of design and art, thinking and practice.

NOTES

1. The closing Round Table panel was chaired by Hélène Frichot; the speakers were Professors Elizabeth Grierson, Harriet Edquist, and Richard Blythe.

2. See for example, Jacques Derrida, *Writing and Difference*, trans. Alan Bass (London: Routledge, 1995); hereafter cited as *WD*. See also Jacques Derrida, *Margins of Philosophy*, trans. Alan Bass (Chicago: University of Chicago Press, 1982). Hereafter cited as *MP*.

3. With respect to the argument about post-critical theory, or the death of theory in architecture see: Michael Speaks, "Which Way Avant Garde," in *Assemblage*, no. 41 (April 2000); Speaks, "After Theory," in *Architectural Record* 6 (2005); Speaks, "Intelligence After Theory," in *Network Practices: New Strategies in Architecture and Design*, ed. Anthony Burke and Therese Tierney (New York: Princeton Architectural Press, 2007); Sarah Whiting and R. E. Somel, "Notes around the Doppler Effect and Other Moods of Modernism," in *Perspecta* 33, (2002); Whiting and Somol, guest editors, *Log* 5 (2005); see also Hélène Frichot, "The Death of Architectural Theory and Other Spectres" in *Design Principles and Practices: An International Journal* 3 (2009), http://www.design-journal.com. With respect to "theory trouble" in the fields of literary theory, comparative literature, and philosophy, see also *Symploke: Theory Trouble* 11, nos. 1–2 (2003).

I

De-signing Design

Where Lies the Art of It?

Elizabeth Grierson

What might *de-signing* mean in context of spatial systems, and the informational and social dynamics of urban spaces? What kinds of trajectories are at play in processes of designing when design couples with architecture and gestures towards imperialist discourses of urban place-making? Is it possible to intervene in design to excavate a critical process or way of thinking, to go some way towards disrupting customary ways of thinking and doing, or to de-sign "presence"? The aim here is to move in the spaces between theory and practice, to prise open discourses of design while opening differentiated sites of practice for examination. With this in mind, the following chapters activate ways of questioning pre-assumed ideas of art, design, place-making, mapping, subjectivity, and appearance.

Part I of this book evidences acts of putting theory to work in the critical exigencies of practice as the five authors find ways to put understandings of creative knowledge and design to the test. This is not merely a theoretical exercise, but a serious way to grapple with a mode of responsibility by engaging a radical politics in design practice, thinking and action. Elizabeth Grierson, Scott McQuire, Mark Jackson, Marsha Berry, and Maria O'Connor bring a critical lens and differentiated positions to the theme of *De-signing Design*. Each writer brings to the fore their particular focus on diverse theoretical and practical underpinnings of design, space and place, laboring, being, and belonging. Much of this work identifies ways of reading the conditions of appearance with a sense of spatiality in one's social dimensions, memories and experiences. As a group of chapters there is consideration of the tropes of creativity, knowledge and design with particular attention to the ways they frame and form our understandings of the city—the city as it was, is and may be, and ourselves within it.

This introduction, "De-signing: Where Lies the Art of It?" asks why there might be an interest in de-signing design and what this may mean in terms of an artful strategy of place-making. There is in evidence a process of questioning and dismantling normalizing assumptions, and seeking other ways of disclosing, reading, and knowing through design. The writers grapple with ways of deconstructing or reading design in ways other than the productionist discourses of economic knowledge exchange. These texts reveal philosophical, critical, and personal approaches with each presenting different ways of mapping in the cartographies of theory and practice.

In considering the notion of de-signing within discourses of innovation, there is a radical place for a political—and poetical—sensibility. There is a necessity to work towards de-signing because the dominant discourses are becoming increasingly "naturalized" and thus emptied of any possibility other than that of meeting ever-present market needs. With the economization of knowledge in designer-capitalism as the defining mode of twenty-first century exchange, there is an exponential growth of cities and the instrumentalised adoption of creativity and innovation for economic and progressive ends. Through this process, creative knowledge and design take center stage in popular and political discourses as a way of identifying, positioning, and "reading" the city of the twenty-first century and situating the entrepreneurial self within those spatial dimensions. Moving from the late twentieth century when self-referential design statements in architectural works held their authority in a hierarchy of value and meaning making, today new digital dimensions open possibilities of distributed knowledge, authorship, and readership in designed spaces. In context of these shifts, there are changes in the ways producers and consumers might think about the production of urban space and a city's economic, cultural, and social life. Designer cultures carry the weight of economic expectation in urban planning and development with universities as knowledge-producing institutions folded closely into the frames of strategic development of public spaces. Universities seek to work with industry in financial and knowledge-based relationships, as "the urban" figures more strongly as a priority field of research—and possibly funding. Universities partner with industry to form knowledge-hubs for the production of utilitarian knowledge and its transfer. These changes in institutional priorities have opened relevant sites of investigation.

The five chapters of "De-signing: Where Lies the Art of It?" examine philosophical, political, and personal dimensions of "de-signing" with particular attention to the devices of language, art and design, creativity, poetics, narrative, and technology within the urban context. They consider theoretical and philosophical ways of raising questions to do with imperialist discourses of design in city planning, architectural theory and practice, laboring and the human subject, memory, and place-making. They demonstrate in a range of ways how there may be interpenetrations of different types of knowing and being, both in and of the city, of selves and of "populations." The authors are seeking different ways of talking about, interrogating, and de-centring commonly held assumptions in and of design. They are dismantling customary or "proper" signage of design, and opening design, as a way of thinking and doing, to its supplementary possibilities. Displacing closure of meaning their writings give rise to what Maria O'Connor names as "an ethics of a self that is radically opened by the other within us." If this ethics, and this self, are not determinable, they may be in differentiated states of becoming, seeking to understand conditionality as a way of signing an "otherwise" position.

Part I starts with Elizabeth Grierson's chapter "De-signing the City: Interventions through Art," which seeks to show how we may work in the spaces between theory and practice to displace over-rationalized and binarized systems that so pervade productionist thinking. Seeking to sign the city otherwise, Grierson works through writings of Jacques Derrida to deconstruct and rupture dominant discourses of design in city planning and place-making, while at the same time grappling with the problematic of her task: "The writer here is interested in de-signing this process, but if words are signs, and signs presuppose positions and meanings, then I am not sure that de-signing is possible," she writes acknowledging this conundrum. She seeks a way of deconstructing these presuppositions in discourses of mechanistic and productionist thinking in space and place-making asking if it is possible to perform a de-signed creativity to sign the city otherwise.

By engaging with Jacques Derrida's deconstruction, the aim is to become more aware of mapping, through difference, our critical histories of the present. This acts as a way of performing an active and creatively aware engagement with the discursive practices of deferral and undecidability in the emergence of innovative thinking, which, she claims, is packaged too easily at the service of urban and global economies of designer capitalism, losing thereby its deconstructive potential. In this endeavour, Grierson's writing also seeks to reinvigorate an understanding of siting and situating creativity beyond the metaphysical laws of unity, propositional logic and self-presence. If creativity is to have hopeful space to breathe as an element in play then we must take it seriously. Through processes of seeking other ways of seeing and inscribing life, work, action, and relationships in these entrepreneurial times, Grierson is bringing a sharper focus and understanding to present horizons of living in urban contexts as a creative, and no less political, way of being.

In "Toward De-signing: Narratives, Networks and the Open Work," Scott McQuire examines the role of narrative in processes of the built environment that comprise contemporary urbanism. He interrogates architecture as a textual construction, a way of signing and regulating order and sequencing, by referencing the ways distributed interactive networks disrupt coherent narratives of sense-making that have so dominated the constructions of architectural theories and the histories of place-making. With user-generated content, participatory media, and peer-to-peer networks, there is a shift in the concept of agency and authorship, as text itself becomes less stable with the disruption of logical sequencing of narrative. McQuire's main concern is to excavate the modes of practice and understandings of architectural narrative and bring them into close proximity with the way users or audiences experience space in a mutating set of conditions. He brings digital modes of production in interactive media and networking, with their destabilizing tendencies, into a questioning relationship with the ordering of narrative in architectural terms. Putting authorship into doubt, the active role of reader takes prominence—a situation that assumes a cultural and political significance with the unpredictability of meaning in information transfer. Umberto Eco's "open work" provides the framework to extend this discussion to the possibilities of distributed authorship as a process of collaborative cultural production of meaning in and of urban spaces. This contributes to a participatory urbanism and the potential for a displacement of instrumental mastery over city planning and city life. His ultimate aim is to posit new forms of "open" architectural narratives involving radical shifts in thinking, a letting-go of previous assumptions about authority and control, in the mobilization of new negotiations and relationships in the spaces between author-architect and user-participants to generate what he calls "unpredictable social alignments."

In "Designations," Mark Jackson seeks a re-design of design engaging across key texts to see design not as a series of acts, but rather as something outside the agency of a laboring human subject. He engages closely with the proposition that the city is a juridico-legal mechanism at the level of governed territory and populations, a spatial mechanism of design as a technology of power, and a site for risk, probability and normativity. He draws from the work of Michel Foucault to trace a genealogy of two discursive fields in the eighteenth century: disciplinary mechanisms and apparatuses of security as ways of ordering the problematic of "population"; and he brings the questions of relations between individuals and population into proximity with the twentieth century, through the writings of Walter Benjamin. Jackson positions the potential to excavate the question of design and its re-design via Foucault's governmentality of populations, which he argues is Foucault's way of addressing conditions of possibility for concrete practices of design. In the relations of discourses, technologies, and practices, the modalities of power are manifest in the ordering of this category, "population." Jackson is concerned also to excavate modernity's legacy of design and couple his excavations of the designed ordering of the populace, with an investigation of the emergence of aesthetics' autonomy in the nineteenth century and its discursive dispersion across a range of fields. His turn to Benjamin brings some proximity to Martin Heidegger, to reveal a way of overcoming the metaphysical impasse of Western history's dualistic thinking on essence and appearance, science and aesthetics. Methodologically, Jackson folds his excavations of Foucault, Heidegger, Kant, Hegel, and Benjamin into a genealogical text that moves the potential for design beyond the laboring self into a receptive openness.

From Jackson's philosophical enquiries, there is a shift in tone to investigations of the effect of postmemory on specific sites of cities and human subjects in the palimpsest of time and experience. In "Signs of Postmemory in Dresden: Restoring the Displaced," Marsha Berry narrates her place-making through a process of postmemory in which she engages the designs of Dresden as a place devastated in World War Two, a place of atrocities, a home from which families were exiled—her family, to be specific, and through postmemory, herself as writer and observer. The process of place-making is overlaid with survivor memories etched into the next generation's experience of time and place—then and now. Berry reads Dresden against monumental historical events of World War Two and layered glimpses of remembered personal experience of exiles and displaced persons through photographs and stories, the pain of forgetting and re-recognition—this a multilayered history of time and place accessible only through words and images of significant others. Cities, people, and places of another's past become more real than the present reality itself, as images in old photograph albums belie their powers of persuasion to texture a childhood far from the monumental events that gave rise to their haunting quality. Working through Marianne Hirsch on postmemory, Berry reveals how the fabrications of vicarious tales of local knowledge of another time may filter the actuality of daily events in the here-and-now time. Atrocities and domestic themes find their way into the present to become proxy identities for a second generation, creating different relations with past experiences, another's past. Recognition is bound to a world of tension, another's exile, a diaspora, a forgetting, a displacement, and a remembering in a shared cultural experience of place. There is something raw in the statement, "Dresden was burning. The awful beauty burned itself into her memory," as the writer draws reader into close proximity with the firestorms and advancing Red Army through travail and profound loss. Dresden in present-time reveals a different place: an estrangement occurs, when there is a mismatch of postmemory, memory, and present reality with detached referents opening spaces of absence. Yet in the gaps there may be a plenitude of possibility. Places contain ghostly traces of traumatic events, which echo through creative practices.

Berry's poetic narrative opens the terrain to Maria O'Connor's "Posed Solitude: Signing a Poetics of Community." Here a performative writing across texts excavates another way of finding a poetic—in community. But, asks O'Connor, how can such a community exist? She seeks to activate this question by engaging closely with the non-place, or *posed-solitude*, of Marc Augé's work, the politico-ethical poetics of Maurice Blanchot's *unavowable community*, and the *ecstatic temporality* of Martin Heidegger. What is crucial here is an understanding of community as something other than the bonding of shared familiarity in the ordered and pre-planned spaces of urban settings. Today, urban planning and spatial designing leans toward the structuring of community precincts with the envisaging of certain behaviors and practices, and the discouraging of others—such is the signing of community as a static or normalizd condition. O'Connor seeks to interrogate this condition and take us beyond into a poetics for de-signing urban living by way of imagining "an otherwise system" with a Blanchotian desire in action. She speaks of "a radical drift" across her texts as she opens the possibilities of engagement with de-signing as a way of problematizng fixity and closure. In activating the notion of community as a new form of spatial practice, she seeks an inventive and interventionist schema and works through crucial aspects of Heidegger's *Dasein* to posit something both within and outside of time, dismantling binary thinking and placing on us a demand or responsibility beyond mastery. She takes the reader into the underworld with Orpheus and Eurydice where Orpheus transgressed the Law through the movement of desire. Here is Blanchot's excess and the excessive zone of rehearsal where the self is possessed by the gaze of the other—that which exceeds us and sources the poetic desire. Here is Blanchot's community, his desire of worklessness as the "other night" that O'Connor describes as "a dying stronger than death in our self, riveted to existence." The way the content of this chapter unfolds displaces metaphysical assumptions of presence as the writer performs her writing. Perhaps there is an absence in the de-signing process? The reader may experience a profound sense of the poetic through an opening to a possible community beyond any previously thought.

Throughout these chapters the writers design possible entries into new forms of thought. Their words disturb pre-panned assumptions and open possibilities of design to new ways of signing. Here lies the enactment of new cartographies of theory and practice. The art of it may reside in the interweaving of philosophical texts with activations of critical practice. Each writer works with different methodologies as they interrogate and evoke their subjects. In so doing, a textual weave destabilizes design mastery. New ways of being in the world become available by de-signing and exceeding the limits of what, once, might have seemed the possible.

Chapter One

De-signing the City

Interventions through Art

Elizabeth Grierson

The following text undertakes a creative investigation of signing, mapping and negotiating urban design and human habitation. Today, discourses of innovation and the entrepreneurial subject embed the concept of creativity as creative knowledge and design take center stage in popular and political discourses of the twenty-first-century city. This chapter places the subject of design within tropes of creativity, knowledge, and subjectivity with particular attention to ways these discourses frame and form our understandings of the city—the city as it was, is and may be, and selves within it.

Negotiations of space, place and time call for attention. Knowledge production and its transfer has become a hallmark and impetus of urban growth, and the equation of "innovation" with "creativity" is a driving condition for the ends of globalized economic progress. This utilitarian view is open for interrogation. The aim is to open modes of signing and place-making to critical enquiry. This chapter works with methodologies from Jacques Derrida, namely deconstruction, as a way of de-centering dominant practices of thinking, writing, and designing. It thereby unpacks the notion of signing with particular attention to the devices of art, design, and creativity within urban contexts to sign the city otherwise.

STARTING POINTS: A CONUNDRUM

"—Sorry, but more than one, it is always necessary to be more than one in order to speak, several voices are necessary for that. . . ."[1] The liberal subject of reason thinks in terms of singularity as a form of centrality, but cities are multiple. They speak of many-layered narratives with cacophonies of sounds, sights, voices, languages, regulatory devices, spatial ambiguities, proximities and identifications, and always the ubiquitous sign systems. The structure of signs becomes a dominant way of reading spatially and mapping the way until one hardly sees the structure itself. It appears natural as an accustomed cartographic language and, with an "impudent fashion,"[2] it presents a dominant syntax of urban dwelling marking the terrain of our days. This discussion is working with the system of signing to open it for interrogation. By tracing genealogies of signing it engages with Jacques Derrida in de-centering the structure of signs, and also brings forth questions from Michel Foucault. The aim is to unsettle normative meanings of design by bringing a sharper focus to present horizons of time and being.

While city planners strategize and promote "good design" as a way forward to progressive futures, artists and designers may seek ways to intervene in such identifications through trans-disciplinary moves. Such interventions may rupture the smooth surfaces of normative planning discourses until different ways of signing cities, design, and selves as human subjects becomes apparent.

The chapter title, "De-signing the City: Interventions through Art" evidences the first sign of a narrative. These words, separately and together, stand for something, name something; they sign something. They act as the nominal starting point in a cartographic exercise of writing the city. The aim is to write freely as a way of thinking in an unconstrained way, but the problem with titles is that the words stand there as engravings claiming a way of representing a position. The words stand there directing the passer-by through the exigencies of text and intention. The writer here is interested in de-signing this process, but if words are signs, and signs presuppose positions and meanings, then I am not sure that de-signing is possible. As soon as one adopts a position there seems to be a normative process of signing going on, and if one moves away from that position one soon finds oneself signing again. The possibility of de-signing seems to be ever diminishing. Yet, I will be facing this conundrum, as I seek a way of intervening in dominant discourses of productionist thinking to see if it is possible to perform a de-signed creativity to sign the city otherwise.

MAPPING, SIGNING, LANGUAGING: DE-SIGNING THROUGH DECONSTRUCTION

The term de-signing implies a form of deconstruction, so perhaps Jacques Derrida will assist here. For Derrida signs achieve their meanings from other signs in a chain of reference or deferred signifiers. In *Of Grammatology*[3] Derrida shows how the sign is already a structure of difference. It comprises *traces* or *tracks* of others, and deconstructs the notion of presence embedded in the centrality of the transcendental signifier. In his thesis on deconstruction, Derrida puts the sign under erasure. Is this what one must do to move between the discourses of theory and practice in seeking a practical and creative way of mapping, signing, and languaging? Derrida shows how the sign both differs from its other and defers to its other, at one and the same time. For this process, Derrida coined the term *différance*. This explanation may be useful:

> Derrida's deconstruction[4] of metaphysical presuppositions of structuralism may provide conceptual (poststructuralist) procedures for interrogating visual images and objects as texts, with the *deferral* of possible locations of meaning from sites of the signified to signifier, then from signifier to signifier through processes of *deferral*. Derrida works through strategies of *decentering* the text, via *deferral* of what is assumed to be the center of "presence" in structuralist thought and metaphysics; the sites of possible meaning deferring from signified to signifier, and from signifier to signifier (*POK*, 152).

If there is a constant deferral from one sign to another, one signifier to another, then how does mapping work as a series of signs, notations on a flat surface through which and by which navigations occur in the traversal of cities? Ways of identifying place in the world promote customary projections of land on two-dimensional surfaces. They map and mark the juxtapositions of shapes and dimensions of geographic and political territories. A genealogy of mapping, from sixteenth century pictorial depictions of space to the computer generated web systems of today, reveals the consistent human need for instructional devices in the orientation processes of way-finding.

SIGNING BY MAPPING

In 1570 Dutchman, Abraham Ortellus produced the first atlas *Theatrum Orbis Terrarum*, exhibiting the land in the way a theatre displays characters and events. Following those early pictorial displays, map makers represented the world as a collection of territories through a range of projections marked by grid lines of latitude and longitude oriented to a flat surface. Cylindrical, conical, and azimuthal projections present landmasses, seas, and islands of the globe as arrangements designed with mathematical accuracy. By the way onc reads those projections there is a search for exact correspondence of spatial configurations as, perhaps, a human need for unity. Yet each projection reveals a different visual map with differing sets of decoy shapes and relationships aiding the comprehension of territories by their relative sizes and proximities. Signs of land, or "truth" to land actually start to appear as assemblages of multiple possibilities projected as cartographies of mediated construction that decry any unified reading. What seemed a natural representation, or truthful depiction of spatial arrangements, is in fact a cultural artefact activating diverse possibilities of seeing and perceiving.[5]

In translating this situation to processes of mapping a city, one can witness the activation of two concurrent conditions: the linear narratives of city spaces signed by maps and instructional designs as orientation devices, and the interventions of embodied practice as a way of occupying those spaces. Beyond the singular plane of significance as indicated by the diagrammatic designs of two-dimensional maps, actual human practices, and spatial navigations offer different dimensions of perception and articulation. Multi-facetted relations between human subjects and social spaces become subjective ways of encountering the world. Memory too plays a part. One is here and there at one and the same time; with layers, echoes and voices in the present, the remembered past and possible futures. Thus, singularity escapes presence, as several voices impose and interact. Those voices may be ready-to-hand for putting ourselves to work, or may inhabit and activate the temporality of memory and imagination. Memory tells us there was a shop on this corner or we met a friend here or experienced danger there. These memory notations become spatial signifiers as haptic and cognitive processes reverberate and intermingle.

In working with Jacques Derrida the aim is to show a way of moving with the events of practice in the structure of cities—events as "rupture" or "break" moving us away from the expected and customary centrality of the logos, propositional thought and the totalisation of presence. In other words, it is a way of working with the *logic of difference* or multiplicity rather than the logic of singularity. Today the corporatization of city planning proclaims futurist visions of "client service delivery" bringing singular purpose into the economy of design language and practice. Strategic planning missions equate success with value-adding initiatives to establish the sustainable design and production of public projects—"We do what we promise and deliver exceptional economic, social and environmental outcomes."[6] Here appeal to a normative social ethic comes into proximity with pre-figured claims of "best practice"—even "exceptional practice" taking it beyond best. Such statements position productionist development as a progressive feature of "excellent" urban growth.

The language proclaims an economic self-presence as a *gate-keeping* proposition, a standards-setting rhetoric following a self-proclaimed, norm-referenced, common sense approach in which clients or consumers must put their faith. This, the language of a present-day sustainability phenomenon, speaking the rhetoric of good design for social and economic benefit as a neo-liberal norm determined by appeal to metanarratives of progress and reason. Such promises act as a predetermined judgement, so-to-speak, a way of signing social and cultural values as it structures economic standards of production and exchange. That it is a constructed response to a political set of conditions is obfuscated; so too is the ideological inscription of neo-liberal values and attitudes. Derrida's deconstruction of structure may offer a way of dismantling such assumptions by opening them for question; there may be ways of rupturing normative practices to see how discourses of difference may operate as a way of de-signing or signing otherwise.

DECONSTRUCTION

Derrida's project is one of deconstruction operating theoretically and practically. Thinking through deconstruction it is helpful to consider its placement in a history of Western philosophy:

> When Derrida deconstructed philosophy he worked within the discourses of philosophy not outside of them. In his seminal conference on Structuralism at John Hopkins University, Derrida (1966) first critiqued "structure" as inhibitor of the play of meaning in language and text. His methodology was not of the Hegelian dialectic, but of working within the logic of difference. Derrida shows the logic of difference not as contradiction that requires resolving, but as a productive site of deferral. This is a way of reasoning that allows for difference to remain as difference; and for this Derrida coined the term *différance* to gesture at the way words and forms differ from each other and defer one to the other in the search for meaning. In a lineage of practice from Nietzsche, Derrida was putting philosophy to work in new ways, activating philosophy as a living discourse rather than a foundational set of principles (*CAR*, 151).

Derrida offers something here by way of counter-readings of the fundamentals of the coherent linguistic and philosophical structures through which we come to know the world and ourselves within it. He offers a way of understanding discourses as signs already constituted in difference, thereby seeing signs differently while working within their structuring systems. In "Structure, Sign, and Play in the Discourse of the Human Sciences," Derrida points out that structure "has always been neutralized or reduced, and this by a process of giving it a center or referring it to a point of presence, a fixed origin . . . to orient, balance, and organize the structure" (*WD*, 278–94). It is difficult if not impossible to speak of structure without structure, but Derrida's concern is to show how the organizing principles of structure limit the "freeplay" potentials of the structure and therefore of practice. This implicates creativity as a way of thinking and acting as it suggests a way of performing a free relationship with practice.

Any complex argument has its signposts along the way of constituting meaning. The following words act as one of those signposts: "The center is at the center of the totality, and yet, since the center does not belong to the totality (is not part of the totality), the totality has its center elsewhere. The center is not the center" (*WD*, 278–94). Derrida draws from Frederich Nietzsche's radical formulation in his critique of the metaphysics of presence and the substitution of interpretation, play, and the sign (otherwise to the sign of "truth" to appearances), and Martin Heidegger's dismantling of metaphysics (otherwise to the sign of "*a priori* being" as pre-given appearance); and there are also traces of Sigmund Freud's critique of consciousness (otherwise to the sign of "unified identity" of the human subject). Importantly, Derrida notes we cannot dismantle or deconstruct the organising principles of philosophy by presuming oneself to be outside of those same principles. One is implicated, already and totally, in the words one speaks, in the very terms with which one seeks to contest.

So it is with design. Already implicated in the discourse of signing and meaning-making perhaps the task is to seek a way to problematise any assumptions and neutralizations of meaning that make up the discourse. "To "deconstruct" a text is to draw out conflicting logics of sense and implication, with the object of showing that the text never exactly means what it says or says what it means" (*POK*, 152). Deconstruction involves a process of what Christopher Norris calls a "rhetorical close-reading that seizes upon those moments when philosophy attempts—and signally fails—to efface all knowledge of this figural drift" (*POK*, 152).

CREATIVITY FOR INNOVATIVE CITIES

Thus, living in and through language implicates the speaker, writer, and designer in the structural terms of any text. This includes the discursive social and political texts (practices) of globalized knowledge economies with their fast transfer of ideas, information and capital. The globalized economies led by post-capitalist nations proclaim the fundamental importance of creativity for the making of the entrepreneurial city and innovative nation of the twenty-first century. Creativity and creative enterprise are positioned politically, seemingly naturally, as key economic drivers. In this scenario how is the creative subject determined, and where lies the figural drift of the subject in the networked discourses of art, design, and the public sphere? From the propositional assumption that globalized knowledge networks are "creative" by definition and operation, the following question ought to be raised: Is it possible to sign creativity otherwise, to see creativity beyond or outside the metaphysical laws of propositional logic, unity, and self-presence that are structuring our economic futures?

Creativity, from Latin "*creare*," to grow, to bring (something) into existence, gives rise to the concept of creativity as a state or process of flux, change, transformation, of making (something) original, or rearranging certain conditions to revisit, renew, or reinvent knowledge, to rethink the center or law-making structures, add to them or think them otherwise. In Derridean terms, "play" constitutes a way of de-centering the centrality of structuring. In the terms of American sociological writer Richard Florida, creativity in the economic development of cities and regions is a kind of revitalization model of social and cultural life. Reinforcing the kind of popular thinking that underscores economies of creativity, Florida opens *Cities and the Creative Class* with this premise, "Cities are cauldrons of creativity" (*CCC*, 1). The proposition is clearly stated. However, the creativity hovering around the field of design, or at least the *potential* of the field, is the very creativity that is by-passed too easily in the lighting of these economic cauldrons. On the other hand, the creativity of difference and workings of *différance* can activate possibilities of de-signing design, as new awareness of creativity is put into play. How may art figure in this process?

A creative form of *différance* in action was evident in a public site at quayside locations of Newcastle and Gateshead, in 2005, when 1,700 people volunteered to gather naked for the American photographer Spencer Tunick.[7] The artist used the massed naked human form as a way of bringing attention to the urban spaces in which humans dwell. The human form *en masse*, became a site of rupture. The familiar urban locale with its regulatory systems of signs and customary vistas of roads, buildings, bridges experienced a disruption. The mass of flesh in a public space served to displace the familiar or conventional ways of reading spatial relationships to perform a reversal or creative incision in the aura of neutrality that seeing assumes. This site-specific artwork performed both a deferring to, and a differing from the stark geometries of the constructed environment of brick, concrete, glass, and steel. Each came into proximity with the other. Through the activation of time in relation to space, the formation of 1,700 naked bodies had a confronting immediacy for participants and viewers, disrupting familiar social mores. The performative event triggered questions about art and aesthetics and their playful, and serious, roles in spatial and ethical perceptions of public and private space. It triggered questions about laws and regulations. In the deferring and differing processes a new awareness occurred—in both corporeal and structured environments. Tunick's aim might have been to reveal first and foremost the built environment of the city quaysides rather than that of human form. Yet, by shaking up the structure of customary ways of seeing and experiencing designed environments (built and human), each brought the contingency of the other into view. In terms of the structure of signs as orientation devices, the sign both differed from its other, and deferred to its other, at one and the same time via a de-signing process.

De-centering via deferring and differing is evident also through the filmic work *Garden*, 2007, by Chinese artist, Chen Qiulin shown at the 6th Asia Pacific Triennial of Contemporary Art in Brisbane, 2010.[8] From daybreak to dusk, a group of migrant workers wind their way through stacked housing developments to deliver giant ceramic vases filled with fake peonies to the artist's hometown of Wanzhou in Sichuan Province, China. This area was mapped already by its destiny: marked to disappear, "to be submerged by the rising water of the Three Gorges Dam . . . the world's largest hydro-electricity project" (*SR*, 75). Against a background of disenfranchisement, displacement, human rights violations, resettlement problems, fast development and spiraling costs, the pink and red flowers bobble along playfully. The messengers balance the large ceramic vases against their bodies, carrying them at a steady pace along narrow paths, up and down stone stairways, threading through endless, grey concrete structures. The peonies in Chinese art signify "the fragility of life and its potential for renewal" (*SR*, note 3). Here, in the activation of *différance* the poetic lightness of the flowers plays with this potential, thereby drawing attention to the gravitas of political and social upheaval that traces through the city's habitation. The fragmentary movements of pink and red blooms dance provisionally with light, and without agency, in counterpoint to the workers with their resolute footprints threading their way through the haze and detritus of a polluted urban landscape.

KNOWLEDGE AND SUBJECTIVITY

The above encounters with art as artful design reveal the ways art may intervene in what Derrida calls "the motif of presence" (*G*, 97), displacing or de-signing the centrality of this motif, which Edward Soja conceptualizes as the triad of space, time, and social being.[9] There is a disruption to the normative political, social, or spatialized structures of urban design as a causal set of relations. Through the non-agentic dance of the peonies, or the unexpected sight of naked bodies inhabiting an otherwise regulated public space, differentiated elements are put into play with one another.

The shift in the discourse is not about dissipating normative understandings of cities and urban design. Rather, it is to sharpen discernment through activating traces of other perceptions and cognitions beyond liberal tenets of the unified subject, and the utilitarian expectations of neo-liberalism, thereby shaking up any notion of fixed or pre-determined outcomes. There is a displacement of centrality in the structure itself: a demonstrable process of de-centering signs of economic reason and progress. New awareness of both the city and selves as human subjects is taking place. The human subject as a spatial, temporal, and social being is constituted not as a singular entity in a set of symbolic norms and forms; rather it suggests a series of de-centered practices, always in process, provisional always. Michel Foucault enters the conversation here:

> So it is not enough to say that the subject is constituted in a symbolic system. It is not just in the play of the symbolic that the subject is constituted. It is constituted in real practices—historically analysable practices. There is a technology of the constitution of the self, which cuts across symbolic systems while using them.[10]

One of the points for consideration here is the becoming of a human subject in the contingent discourses—and technologies—of our time, space and place, with ineffable traces of other times, other places and other social relations. It seems that place, particularly urban place, situates a discourse of complex relationships via the logic of difference in material practices, technologies, systems, social patterns, representations, assumed certainties, and the uncertainties of misrecognition and misrepresentation. Within discursive and fragmentary practices the human being is always "in process" becoming a human subject in the temporal practices of thinking and doing.

In this process, identifying oneself in and through one's locale as a communing and communicating subject, one is not a singular being; one is, as a "being-with," constituted in difference. There is a multiplicity in the one who thinks and acts, responds and identifies, doubts and clarifies while navigating everyday realities. As one walks, runs, cycles, drives, trams, or buses around a city, processes of place-making activate the navigations of differentiated exchanges, interiorized and exteriorized, not always rationalized or logically determined. Through poststructuralist theories and methodologies one may apprehend the displacing and de-centering of the *cogito*, as a multiple human subject registers difference-in-action: forming, deforming, transforming in time, never having arrived. A differentiated process of knowing and being with reference to Michel Foucault works like this:

> . . . there is renewed attention on the subject as a process of becoming. Foucault is concerned with how a human being transforms him or herself into a subject. . . . This moves us away from the Western Enlightenment narrative of the progress of an a priori human subject, already established in its essence through the cause of reason, and coursing through history with the pre-set goal of transcendence of the spirit to a utopian endpoint (*CAR,* 42).

UTOPIANISM

Liberal discourses of utopianism have a history. It could be said that utopianism, underlying the liberal metanarrative of progress, holds a powerful political presence in urban planning and political thinking about the human subject. Such a presence is, however, assumed and undeclared, busily setting up means-end relationships with the hoped-for future.

A pause to consider discourses of utopia reveals infusions of natural law or common sense constructions through claims of personal autonomy in the advancement of a fully rational human subject. Existing in the space of imagination, *utopia*, "no place" (Greek "*ou*" not, and "*topos*" place) claims an undisputed position in the histories of liberal reason and scientific expansion through projects of Western Enlightenment thinking and action. In suggesting that discourses of utopia are available for intervening and de-signing, then it would seem appropriate to trace and disclose what Michel Foucault refers to as, "the modes of existence of this discourse."[11] Foucault asks, "Where does it come from; how is it circulated; who controls it?" (*MFL*, 138). Tracing a discourse of utopia takes us inevitably to tropes of transcendentalism:[12]

> The term was first used by Sir Thomas More in 1516 as a Christianised concept in his book *Utopia*, an imagined place, a state of things where perfection may be possible . . . signifying hope in belief beyond the confines of physical place, its opposite "dystopia," an imagined place or condition of perfect *im*perfection. In the western world the tradition of liberal education, appealing to the highest good of reason, was understood as a means towards a utopian adulthood, that time of becoming a fully rational human being. [G]uided by the dictates of reason . . . the adult leads the child learner along the path of righteous and responsible morality towards the achievement of autonomy and objectivity. Reason reigns (*FF*, 7–14).

The twenty-first century reignites utopian desires through economic reason in the creative economies of globalization. The implication is that if the laboring self is at the service of the economy then all will prosper and one's productive and moral relationships to the world at large, and to time, will be more ethical, efficient, and effective. Producers and consumers witness the transformation of pastoral, industrial, and post-industrial economies into smart economies based on knowledge production, fast transfer and exchange of information, and the ever-increasing technologisation of social spaces in an age of post-capitalism—the new utopia of a technologized twenty-first century proclaiming itself in many familiar ways.

In the transcendent interests of creative economies as the new discourse of economic reason, city planners look to sensible design as a way forward for the urban locale—this, the utopian future. In coexistence, artists and designers may seek to intervene and rupture the privileging of utopian thinking. Through questioning and displacing, they may initiate critical ways of identifying where the creativity of knowledge may lie. They may ask how a way of de-signing dominant structures of the sign may intervene in urban lives. In poststructuralist thinking such interventions can be called counter-readings, "posing a critique of propositional meaning and disrupting history's heavy emphasis on the unitary autonomy of the cognitive self, coursing through history as an agent of progressive betterment of the civilised condition" (*CAR*, 161).

QUESTIONING THE TERMS OF ENGAGEMENT: CONCLUDING WITH DE-SIGNING

This discussion started by recognizing that titles act as signs to inscribe a form of centrality in the structures of place, space, and being. It is concluding in a relationship with the same sign, albeit from an inevitable distancing. This move is serving as a final reflection on how art and design may activate and intervene in urban place-making, and how thinking about art and design in other than productionist ways may draw together a constellation of creative responses in urban discourses.

Design can so easily become an imperialist discourse in urban place-making. It is customary for property developers to work with architects who commission or select artworks for corporatised spaces. Valued for their object-based qualities of display, artworks at the service of architecture augment rather than activate space—a proposition that places art as a secondary component to architecture, not even as its supplement, but as its add-on. Nietzsche sees beneath the surface of such events of practice. "Such a desire for ownership and unity lies behind the Platonic idea of justice as a 'giving to each his own'" (*NAP*, 81). In following Nietzsche's destruction of normative evaluations, one would see "a philosophical abstraction and figurative concreteness" at work.[13]

By activating deconstruction in designing processes the potential to read and write differently becomes possible. Instead of planners or architects acting as the dominant decision-makers to select or commission artworks for purposes of augmentation, they would be working actively with artists as co-specialists in design, with an enhanced criticality towards the potentials of space and place-making—thus warranting a "playful" relationship through de-signing design processes. As Jane Rendell points out in her argument for critical spatial practice, "works can be positioned in ways that make it possible to question the terms of engagement of the projects themselves."[14]

By questioning foundational terms art has the potential to reveal difference-in-action as shown by Spencer Tunick and Chen Qiulin. Such difference can work also in architectural planning and assemblage. Walking between the cracks of this utilitarian age, with its means-end demands linking creativity to innovation, and design to industry as a primary virtue and function, the *work* of art and the artist may have a significant role to play in the stakes of urban planning and practices. Art may disclose further questions to ask. With a mobility of thought and action, as evidenced by the art events discussed here, art interventions may challenge given boundaries of what the design process can become and art's potential within that process to radicalize a critical thinking and doing. Criticality itself is overlooked too easily in the utilitarian tenets of urban planning and development in an urban age of productionist demands and well-packaged solutions.

This discussion has worked towards a way of de-signing design cultures and their politics through the logic of difference. By revealing a way of signing as *différance*, it opens for question pre-thought assumptions in the structuring principles of thought and action. The aim is to dismantle or de-center dominant discourses and place-making practices in the economy of design—to de-sign the privileged signs, and to let speak other narratives of differing relations and exchanges.

Derrida offers "a kind of writing that implicates institutional authorities as it teaches us to think and act differently."[15] Through prising open sites of signing and designing, of thinking and doing, perhaps some disclosures of our critical histories of the present may become possible.

NOTES

1. Jacques Derrida, *On The Name*, ed. T. Dutoit, trans. D. Wood et al. (Stanford: Stanford University Press, 1995), 53. Hereafter cited as *OTN*.

2. Friedrich Nietzsche, *The Birth of Tragedy*, in Vanessa Lemm, *Nietzsche's Animal Philosophy: Culture Politics and the Animality of the Human Being* (New York: Fordham University Press, 2009). Hereafter cited as *NAP*.

3. Jacques Derrida, *Of Grammatology*, corrected edition, trans. Gayatri Chakravorty Spivak (Baltimore and London: John Hopkins University Press, 1998). Hereafter cited as *G*.

4. Christopher Norris and Andrew Benjamin, *What is Deconstruction?* (London: Academy Editions, 1996). Hereafter cited as *W*. In writing about deconstruction this way I am mindful of what Christopher Norris and Andrew Benjamin have to say about the way the term "deconstruction" is presented, that it would "be wrong . . . to offer a summary account of it as evolving some pre-given sequence of arguments, strategies or moves. For this is to assume—against all the evidence of Derrida's writings—that concepts can exist in an ideal realm of self- identical meaning and value which somehow transcends the contingent fact of their existing in written and textual form. So one can well understand Derrida's impatience with those purveyors of short-cut intellectual fashion who demand to know what deconstruction 'is' how it works or what results it will standardly produce when applied to any text" (*W*, 12).

5. The discussion on mapping is drawn from a catalogue essay. Elizabeth Grierson, "Territories: Contemporary Photographic Work from Australia and China," *Territories* (Project Space, Melbourne, April 2009), accessed January 20, 2010, http://schoolofartgalleries.dsc.rmit.edu.au/PSSR/exhibitions/2009territories/territories.pdf.

6. Statements from "Our Approach" in *Major Projects Victoria* Melbourne, last modified August 16, 2010, http://www.majorprojects.vic.gov.au/about/our-approach2.

7. See Elizabeth Grierson, "Building Dwelling Thinking and Aesthetic Relations in Urban Spaces: A Heideggerian Perspective on Relational Pedagogy as a Form of Disclosure," in *Aesthetics in Action, ACCESS: Critical Perspective on Communication, Cultural & Policy Studies* 29, no. 1 (2010): 36.

8. Chen Qiulin, "Salvaged from Ruins," in *The 6th Asia Pacific Triennial of Contemporary Art* (Brisbane: Queensland Art Gallery, 2009), 75. Hereafter cited as *SR*.

9. Edward Soja, *Postmodern Geographies: The Reassertion of Space in Social Theory* (London: Verso, 1989); and *Thirdspace: Expanding the Geographical Imagination* (Oxford: Blackwell, 1996). Soja's conceptual device of trialectical thinking derives from Henri Lefèbvre and is used as a structuring element by Jane Rendell in her analysis of the spatial, the temporal, and the social in "the place between" art and architecture. See Jane Rendell, *Art and Architecture: A Place Between* (London: I. B. Tauris, 2006). Hereafter cited as *AA*.

10. Michel Foucault, 1997, 277, cited in Mark Olssen, *Michel Foucault: Materialism and Education* (Westport, CT and London: Bergin & Garvey, 2006), 153.

11. Michel Foucault, "What is an Author?" in *Language, Counter-memory, Practice: Selected Essays and Interviews*, ed. D. F. Bouchard (Ithaca, NY: Cornell University Press, 1996).

12. This is discussed further in Elizabeth Grierson, "Utopia for All Seasons: Futuring Art—Futuring Education," in *The Chicago Project: Fold-Out Futures* 2, ed. Irene Barberis (Melbourne: RMIT Publishing, 2008), 7–14. Hereafter cited as *FF*.

13. *NAP*, back cover.

14. *AA*, 4.

15. Elizabeth Grierson and Michael A. Peters, "Introduction: The Legacy of Jacques Derrida," in *The Legacy of Jacques Derrida, ACCESS Critical Perspectives on Communication, Cultural & Policy Studies* 24, no. 1 & 2 (2006): 3–14.

Chapter Two

Toward De-signing: Narratives, Networks, and the Open Work

Scott McQuire

If the relation of architecture to narrative has always been complex and contested, it is complicated further by the transformation of narrative in the context of digital culture, marked by the rise of phenomena such as remix culture, user-generated content, participatory media and peer-to-peer (P2P) networks. In this chapter, I will argue that the relation between architecture and narrative *should* be shifting because the social and cultural conditions of narrative are in flux. In order to think through this issue, I draw on an early essay of Umberto Eco's (first published in 1962) where he develops the concept of the "open work." Eco does this primarily in relation to music, literature and visual art, but his arguments remain useful for rethinking the relation between architecture and narrative in the present. By way of introduction, and before turning to Eco, I will make a brief detour through the definitions of narrative offered by Roland Barthes and Mieke Bal.

DEFINING NARRATIVE

For Barthes, narrative is everywhere. In his essay, "Introduction to the Structural Analysis of Narratives" written in 1966, Barthes argues that narrative "begins with the very history of mankind and there nowhere is nor has been a people without narrative [. . .]. Caring nothing for the division between good and bad literature, narrative is international, transhistorical, transcultural: it is simply there, like life itself" (*IMT*, 79). This is clearly an expansive sense of narrative, in keeping with what was then an ascendant structuralist project to identify shared logics across wide swathes of cultural production bridging not only different cultures, but also different sectors of culture. It positions narrative as an essential human attribute at the heart of communication and sense making.

On the other hand, contemporary narratologists such as Mieke Bal have developed a more restricted and formal sense of narrative distinguishing between different layers of information—what Bal dubs *fabula*, story, and narrative text—that collectively constitute a narrative. In Bal's account, the fabula consists of a "series of logically and chronologically related events that are caused or experienced" by characters in a story world (what is often called the diegesis) (*NIT*, 5). A story is a fabula that is perceived from a certain angle, and a narrative text is a particular expression of the story in signs such as words and pictures.

Can a building constitute a "narrative text" in this more specific sense? What purpose is served by positioning artefacts—built structures and urban environments—as a "series of logically and chronologically related events" communicated in signs? Such questions need to be asked from both ends, so to speak, to enable the growing uncertainty as to how we define a story in the present to collide productively with the new exigencies of social experience and subjectivity in contemporary urban space. Bal's concept of narrative is useful here, in that it raises critical issues about agency, causality, and sequencing as constitutive elements of narrative.

I want to use these questions as levers to move some distance from Barthes's assertion of the universal and transhistorical nature of narrative. At a basic level, the effect of so-called "interactive media" is a destabilisation of the fixity of "ordering or sequencing" and the introduction of new dimensions into the dynamics of authorship. As I discuss in the next section, when readers (or the "audience") not only help to select the order of engagement, but assume responsibility for generating some of the content, key assumptions governing the role of the author and the nature of the "text" become less stable. These shifts are complex, but at a minimum they demand greater sensitivity towards the cultural and historical specificity of modes of narrative.

ARCHITECTURE AND THE PROBLEMATIC OF AUTHORSHIP

Let us take the issue of authorship first. If we claim that architectural narrative is about the mobilization of specific architectonic elements and semiotic codes to construct something *like* a story, we are clearly aligning narrative in architecture with the problematic of the author, of the relation of the text to context, and so on. This is a common and arguably dominant understanding. For example, postmodernism, under the auspices of those such as Charles Jencks conceived architectural narrative largely as a reaction to what was claimed to be the repression of history by modernism. "Narrative" became an umbrella term for playful and ironic games of historical reference.[1] This, of course, was a game with a restricted audience, and in this respect, despite often claiming a mantle of populism, much postmodern architecture remained fairly continuous with the *avant-garde* that it ostensibly rejected in terms of its *disconnection* from users.

This highlights a critical issue: when we talk about architecture and narrative is our focus on the author or the audience? What is the place of "public legibility" in relation to architectural narrative? In the opening chapter of his magisterial *The Production of Space*, Henri Lefèbvre discusses the difference between what he calls spatial practices, representations of space, and representational space (*PS*, 33). In Lefèbvre's model, spatial practices enable the reproduction of social relations in a particular spatial location; representations of space were coded forms tied to the dominant order, while representational spaces were complex and less formalized symbolisms often linked to the underground side of social life. These categories related broadly to the dimensions of perceived, conceived and lived space. In terms of enabling citizens to enjoy what he famously called their "right to the city," Lefèbvre argues that these dimensions need to be interconnected, so that "the individual member of a given social group may move from one to another without confusion." He contends that this can only happen in "favourable circumstances, when a common language, a consensus and a code can be established" (*PS*, 40). In other words, there must to be a common narrative shared by creators and users alike—by authors and audiences. He further suggests that while the Western Renaissance town "enjoyed such auspicious conditions" (*PS*, 40), that shared sense of space was "shattered" in modernity (*PS*, 25). Whether or not we accept Lefèbvre's model of a pre-modern common narrative as historical fact, it is undeniable that modernism comes to be characterised by its growing *absence*. This manifests partly in the frequent disjunction between *avant-garde* aspirations and everyday understanding, summarised in the oft-repeated complaint that the public just "don't get it."[2] It manifests also by a growing professional consciousness, and indeed self-consciousness, of this absence.

Toward the close of the twentieth century, the narrative of critical architecture has become less a projection of a coherent future (along the lines of Kenneth Frampton's "critical regionalism") and more a shift to manifestoes underlining the disjunction in which architecture finds itself. According to Peter Eisenman, the exhaustion of all the older orders of reference (symbolic systems, logics of representation, anthropocentric measurement) means that the only critical gesture left to architecture is negativity. In this context, the task of critical architecture is to become a *zero text* in which "the process of the narrative becomes the axis of destruction."[3]

ARCHITECTURE AND THE PROBLEMATIC OF THE AUDIENCE

How might we get out of the impasse of the *shattered* narrative? One possible route is to resituate architectural narrative in relation to the problematic of reception. However, this is neither straightforward nor easy. To return to the earlier point concerning order or sequence as constituent elements of narrative: if the building is a text, how is its order or sequence to be regulated? This has always been an issue for architects. To some extent the control of sequence is achievable structurally through design, but in the end I would argue that such control is less a function of the architect than of the degree to which the audience share codes and protocols of interpretation. In other words, beyond a certain point, it is very difficult to prescribe physically the ways in which people will utilise and move around a building or precinct and this difficulty is magnified vastly if it seeks to encompass the meaning they might attach to their experience.

What Lefèbvre describes as the shattering of shared spatial knowledge is simultaneously liberating and generative of fundamental problems for modernism. How can architecture organize a coherent narrative in a context where not only are the meanings of individual elements (windows, doors, walls, and so on) changing radically, but the broader context of the city is also mutating? How can parts assemble into new orders, which may never become traditional "wholes"? One response was a growing reliance on the system of media, of publicity, to structure architectural narrative. Beatriz Colomina analyses this technique in relation to Le Corbusier, noting the way he explicitly presents his work as an industrial commodity contextualised by the rhetoric and advertising imagery of contemporary products such as automobiles, aeroplanes and electrical turbines.[4]

At another level, she notes the way his designs internalise certain media functions, such as his use of the horizontal window as a frame transforming the exterior landscape into an *image*.[5] Both trajectories are present in Pierre Chenal's 1930 film, *Architecture d'Aujourd'hui*, which locates Corbusier's villas in the context of both moving vehicles and mobile points of view, exemplifying the close relation perceived between architecture and cinema at the time. The widespread desire to render architecture *dynamic* informs Siegfried Giedion's comment about modern design: "Still photography does not capture them clearly. One would have to accompany the eye as it moves: only film can make the new architecture intelligible."[6]

It is instructive to contrast this understanding of the relation between architecture and film with that of Walter Benjamin. Where Corbusier and Giedion imagine the architect controlling perspective and sequence with the facility of a film director, corresponding to a traditional *authorial* perspective, Benjamin compares film and architecture largely in terms of reception. In fact, there are two common elements he stresses: firstly, both film and architecture are consumed collectively, *in public*; and secondly, they are each consumed *absent-mindedly*, in a "distracted" rather than focused state.[7] Insofar as distracted consumption eludes the filters of habit, Benjamin saw the architecture as a reservoir of latent energies and histories, and film as a potentially explosive technique enabling their release.

Interest in the relation between cinema and architecture resurfaced with a vengeance as digital imaging developed through the 1980s. However, it was not Benjamin's problematic of distracted collective reception which came to the fore, but Giedion and Corbusier's problematic of the architect as author—or film-maker. At the risk of over-generalizing, it could be said that architecture in the 1980s and 1990s was supremely uninterested in the Internet as a means of reconfiguring social networks, but was fascinated by the computer as a toolbox for generating sophisticated digital imagery. As computers gained in processing speed and storage, architects were able to build virtual environments using control of perspective, sequence and duration to construct "narratives" in the mode of film directors.[8]

Interestingly, it was around the same time that this model of tightly sequenced linear narrative began to be questioned in cinema and elsewhere. In fact, most filmic experiments in direct user-control proved to be a dead-end, but what did emerge in this period were innovative cultural forms such as computer gaming. Gaming has raised new and critical questions around narrative, particularly concerning the role of the audience in navigating and organizing the order of the text.[9] When users not only find their own pathways but also generate significant amounts of content through their choices and interactions, the old conceit of a single shared text around which "interpretative" arguments circulate necessarily loses ground.

As digital technology has become pervasive, it has contributed to a range of new cultural practices. While it is facile to argue that cultural products ever had entirely "fixed" meanings, digital texts facilitate a new level of "openness" by making textual alteration easier, cheaper, and practiced by more people. It is in this context that we have seen a range of new terms recently emerging to describe the new paradigm of shared authorship: remix culture (Lawrence Lessig), participatory media (Henry Jenkins), post-production (Nicolas Bourriaud), produsage (Axel Bruns), and so on.[10]

THE "OPEN WORK" AND NETWORK CULTURE

To better understand the reconfiguration of narrative in contemporary network culture, I want to turn to Umberto Eco's concept of the "open work."[11] *The Open Work* can be read as an early statement of two themes that subsequently became more prominent in Eco's work and in cultural theory more generally: an insistence on multiplicity, plurality, or polysemy in art and cultural production; and a growing emphasis on the active role of the reader. In his initial essay, Eco sought to clarify the differences between traditional and modern art, arguing that every work has a degree of openness, which is manifested in the question of interpretation. Medieval works, one of Eco's abiding interests, were designed to be read at a number of different levels: literal, moral, allegorical, and anagogical. While this variety necessarily endowed the work with a measure of openness, it was a *restricted* openness hemmed in by protocols shaping interpretation according to a quest or hunt for an exemplary meaning. Eco argues that:

> The meaning of allegorical figures and emblems which the medieval reader is likely to encounter is already prescribed by his encyclopaedias, bestiaries and lapidaries. Any symbolism is objectively defined and organized into a system. Underpinning this poetics of the necessary and univocal is an ordered cosmos, a hierarchy of essences and laws which poetic discourse can clarify at several levels, but which each individual must understand in the only possible way, the one determined by the creative *logos*. The order of a work of art in this period is a mirror of imperial and theocratic society. The laws governing textual interpretation are the laws of an authoritarian regime which guide the individual in his every action, prescribing the ends for him and offering him the means to attain them (*OW*, 5–6).

Eco posits a critical difference in the hermeneutic process of modern culture. For instance, when he discusses Kafka, he argues that:

> Unlike the constructions of medieval allegory, where the superimposed layers of meaning are rigidly prescribed, in Kafka there is no confirmation in an encyclopaedia, no matching paradigm in the cosmos, to provide the key to the symbolism. The work remains inexhaustible insofar as it is "open," because in it an ordered world based on universally acknowledged laws is being replaced by a world based on ambiguity, both in the negative sense that directional centers are missing and in a positive sense, because values and dogma are constantly being placed in question (*OW*, 9).

It is in this context that "meaning" becomes a matter of potentially infinite regress, which Eco will later conceptualize in terms of Charles Sanders Peirce's unlimited semiosis.[12] Eco's model for communication in this context is not a dictionary of one-to-one correspondences but an encyclopedia within which an *infinite* variety of connections can be made and conceptualized in terms of a new vocabulary of the net, the rhizome, or the labyrinth.

Eco then goes further and recognises that certain works embrace openness, and begin to involve the free reorganization of their compositional elements. Drawing on examples such as Alexander Calder's mobiles, Karlheinz Stockhausen's music, James Joyce's novels, and Bertolt Brecht's plays, Eco argues that such "works in movement" are all characterised by the artist's decision to leave the arrangement of some of their constituents either to the public or to chance. This has the effect of giving—or demanding—a greater degree of collaboration from the public. It also heightens the ambiguity of the work, by opening it to different constructions and leaving it "definitively unfinished" in Marcel Duchamp's sense. In Eco's terms: "Every performance *explains* the composition but does not *exhaust* it. Every performance makes the work an actuality, but is itself only complementary to all possible other performances of the work" (*OW*, 15).

RETHINKING ARCHITECTURE AND NARRATIVE

Eco's project belongs to the profound rethinking of knowledge *as information* which occurred in the second half of the twentieth century, marked by the development of cybernetic theory on the one hand, and its counterparts and off-shoots such as systems theory and structuralist semiotics on the other. Eco's originality is to invert the cybernetic model of the relation between information and meaning. Where cybernetics, at least in its early formulations by those such as Norbert Wiener, argued that unambiguous information possessed the highest level of meaning, Eco legitimates modern art precisely in terms of its capacity to convey a higher degree of information because it possesses a high degree of unpredictability.

Drawing on physics, Eco describes the open work signalled by developments in modern art as a "field of possibilities," which assumes cultural and political importance because of its capacity to utilise "*indeterminacy* as a valid stepping stone in the cognitive process" (*OW*, 15). How might we apply this thinking to the field of architecture? What are the implications of the "open work" for thinking about narrative in architecture?

Such questions point beyond the poststructuralist thesis of the "death of the author," which posited a diminished role of authorial intentionality as an interpretative framework, and instead demand consideration of the emergent paradigm of collaborative cultural production that might be called *distributed authorship*. This highlights a new question for contemporary architecture: how might so-called participatory media contribute to something like participatory urbanism? In fact, such a question is not entirely new to architecture. Yona Friedman's *L'architecture Mobile* manifesto of 1958 proposed:

> Constructions should be variable and interchangeable. The spatial units produced by these constructions should likewise be alterable and interchangeable in their use. The inhabitants must be given the opportunity to adapt their dwellings themselves to the needs of the moment (*PM*, 168).[13]

Similarly, the Archigram group's evocation of a "plug-in" city coordinated by computer networks in the 1960s was the vision of a city no longer controlled from above by "experts" such as urban planners, but reconfigured at will from below as the dynamic networked expression of individual desire. In the wake of the extension of digital networks and the emergence of ubiquitous computing these precocious visions assume a new valence. Everywhere you look in contemporary cities, architects, designers, artists, urban planners, advertisers, local governments, and others are expressing interest in how digital networks might constitute informational inputs, in the form of sensors operating on a variety of channels, and outputs, in the form of displays, but also as physical actuators in the environment. The challenge now is twofold. Firstly, to develop and articulate a broader spectrum for thinking about such technological infrastructure outside the still dominant frames of surveillance and spectacle; and second, to design interfaces and programmes which go beyond simple "reactive" responses, or modes of "interactivity," limited to a defined menu, and to aim for more complex goals involving emergent behaviors, adaptation, and learning.

Some of the best examples of developing experimental public interfaces are the various "relational architecture" projects of artist, Rafael Lozano-Hemmer. His recent installation *Pulse Room* (2006) deployed a grid of 100 incandescent light bulbs suspended in a room, activated by a sensor, in the form of a metal sculpture, which was able to register the systolic and diastolic pulses of those who grasped it. The lights flicker in response to this data, mirroring the rhythm and intensity of each person's heartbeat. As new users contribute their own data, the display moves one place along the grid. Eventually, the space is filled with the intricate percussion of 100 different heartbeats forming complex visual rhythms.

An aspect of Lozano-Hemmer's work I admire is the way it reminds us that technology always has more than one mode of use. His installations have frequently deployed surveillance technology, for example. At times, primarily to underline the social fact of surveillance and to show how far it extends into the interstices of everyday life and imagination. At other times, he moves beyond critique to use surveillance technology to generate novel and creative forms of affective sociality, of play and interplay.

Pulse Room performs precisely this kind of displacement. It takes a technique of biometric surveillance most commonly used in medicine, but also routinely applied to security applications, such as polygraphy, and reworks it into a mechanism capable of producing striking visual effects. *Pulse Room* also demonstrates the delicate balance between individual agency and collective manifestation, which is characteristic of Lozano-Hemmer's work. The heartbeat is a quintessentially individual *signature*; even more than a written name (which can be forged), or a photograph (which can be modified), it signifies the inner being of the person. Of course, it is precisely this seemingly irrefutable connection of data and person which underpins the shift in contemporary surveillance techniques to biometrics, as identity photographs give way to DNA samples, cornea scans, and the like. But here the question of authenticity is almost incidental. The achievement of *Pulse Room* is to animate the "secret name" of each pulse into a visible sign. However, instead of being used to *pin* identity to a particular body in the manner of an entomologist, the biometric signature becomes the means by which distinct individuals are woven into a collective tapestry. *Pulse Room* reconfigures the individual members of its audience into a *multitude* in Michael Hardt and Antonio Negri's sense: a temporary collective in which the "common" is not established at the expense of the uniqueness of each member, but instead draws on uniqueness as the integral basis of any relation to others.[14]

Pulse Room exemplifies the logic of Eco's "open work." For Lozano-Hemmer, one of the most important affordances of digital media is the new range of possibilities for "programming without teleology" in which the artist cannot dictate the outcome:

> By means of non-linear mathematics, like cellular automata, probabilistic ramifications, recursive algorithms or chaos strategies, it's possible to write programs whose results will surprise the author. That's to say the machine can have certain autonomy and expression because you simply capture initial "algorithmic conditions" but do not pre-program the outcome. This is for me a gratifying post-humanist message; a message that invites humility, but one that also marks a crisis in authorship and opens a wide problematic area, and I say "welcome" to that! (*SC*, 5).

For an architectural work to be "open" in this sense, it is not simply a matter of "applying" interactive digital technology and "sampling" the audience in some way. Rather, it involves a shift in thinking. Architects need to learn to let go of the belief that they should *control* the narrative and instead learn to accept a new relation involving a process of ongoing negotiation between author-architect and audience as user-participant. There are increasing signs that such a rethinking is occurring in architectural circles, for instance in Lars Spuybroek's notion of "vagueness." Spuybroek criticises the "dry grid" of the classical Miesian box or hall that uses mass production techniques to produce a generalist form of architecture "that can absorb life, chance and change, while the structure itself must last and persist over time, to span the unforeseen with the foreseeable" (*NOX*, 356). While "general openness" may work when all events are pre-programmed (for example, a military barracks), it is fundamentally unsuited to the complex and contingent interactions characterising big city life. The aim of what Spuybroek calls the "wet grid," enabled in part by networked computing, is to displace the "general openness" of modernism with an architecture of "vagueness":

> We must replace the passive flexibility of neutrality with an active flexibility of vagueness. In opposition to neutrality, vagueness operates within a differen-

> tiated field of vectors, of tendencies, that allow for clearly defined goals and habits and for as-yet undetermined actions. . . . It is a structural situationism. It allows for derives and detournements as structural properties: the transparent intentionality of planning and habit is stretched by the sideways steps of opaque intentionality (*NOX*, 357).

A critical aspect of what Spuybroek terms "vagueness" is that it better allows for both formal and informal conduct. Vagueness enables the negotiation of difference. However, if digital infrastructure continues to be oriented toward social relations based on transparency and immediacy, it is unlikely to produce a social space of "vagueness," but one of perpetual customization and hyper-commodification. To realise the promise of Spuybroek's "wet grid" demands a departure from the cybernetic logic of instrumental mastery over the narrative of city life. Such departure is in favor of the design of "unfinished" spaces, which leave room for unplanned, contingent, and unpredictable social alignments.

NOTES

1. See for example Jencks's influential *The Language of Post-Modern Architecture* (London: Academy Editions, 1978).

2. Here I am glossing significant differences between groups such as the Constructivists in the 1920s Soviet Union and the Bauhaus in Weimar Germany, who sought to develop architecture for "the masses," and the explicitly corporate orientation taken by those such as Mies van der Rohe. The basic point is that both groups understand the architect's role as one of authorship and authority over the public.

3. Peter Eisenman, "Text as Zero: or: The Destruction of Narrative," in *Re-Working Eisenman* (London: Academy Editions; Berlin: Ernst & Sohn, 1993), 42.

4. See the chapter "Publicity," in Beatriz Colomina, *Privacy and Publicity: Modern Architecture as Mass Media* (Cambridge, MA: MIT Press, 1994). Hereafter cited as *PP*.

5. Colomina compares the functioning of the windows in Corbusier's villas of the 1920s and 1930s to a camera, arguing: "The house is a system for taking pictures. What determines the nature of the picture is the window . . . if the window is a lens, the house itself is a camera pointed at nature" (*PP*, 311–12). For a more detailed discussion of this issue, see Scott McQuire, *The Media City: Media, Architecture and Urban Space* (London: Sage, 2008), 171–73.

6. Siegfried Giedion cited in Andres Janser, "Only Film Can Make the New Architecture Intelligible! Hans Richter's *Die Neue Wohnung* and the Early Documentary Film on Modern Architecture," in *Cinema and Architecture: Méliès, Mallet-Stevens, Multimedia*, ed. François Penz and Maureen Thomas (London: BFI, 1997), 34.

7. Walter Benjamin, "The Work of Art in the Age of its Technological Reproducibility," in *Selected Writings 4, 1938–40*, ed. H. Eiland and M. W. Jennings, trans. E. Jephcott et al. (Cambridge, MA/London: Belknap Press 2003), 268–69.

8. In this vein Michael Eleftheriades argued that as computers become standard "the world of architecture will merge imperceptibly with the world of cinema." See his "Architecture or Cinema. Digital 3D Design and the World of Multimedia," in *Cinema and Architecture: Méliès, Mallet-Stevens, Multimedia*, ed. François Penz and Maureen Thomas (London: BFI, 1997), 143.

9. See, for example: Espen J. Aarseth, *Cybertext: Perspectives on Ergodic Literature* (Baltimore: Johns Hopkins University Press, 1997); Janet Horowitz Murray, *Hamlet on the Holodeck: The Future of Narrative in Cyberspace* (Cambridge, MA: MIT Press, 1998); *First Per-*

son: New Media as Story, Performance and Game, ed. Noah Wardrip-Fruin and Pat Harrigan (Cambridge, MA: MIT Press, 2004).

10. Lawrence Lessig, *Remix: Making Art and Commerce Thrive in the Hybrid Economy* (New York: Penguin, 2008); Henry Jenkins, *Convergence Culture: Where Old and New Media Collide* (New York: New York University Press, 2006); Axel Bruns, *Blogs, Wikipedia, Second Life and Beyond: From Production to Produsage* (New York: Peter Lang, 2008); Nicolas Bourriaud, *Postproduction: Culture as Screenplay: How Art Reprograms the World*, trans. Jeanine Herman (New York: Lukas & Sternberg, 2002).

11. Umberto Eco, *The Open Work*, trans. Anna Cancogni (Cambridge, MA: Harvard University Press, 1989). Originally published in 1962, *The Open Work* predates Eco's writings on semiotics, although a number of the essays, which make up the English language version (1989), have been revised in light of his later work. Hereafter cited as *OW*.

12. Umberto Eco, *The Role of the Reader: Explorations in the Semiotics of Texts* (Bloomington: Indiana University Press, 1979). Hereafter cited as *RR*.

13. Yona Friedman's manifesto was adapted as the Programme for Groupe d'Etude d'Architecture Mobile (GEAM), which began meeting in late 1957 and lasted until 1962. The Programme is reprinted in *Programmes and Manifestoes on Twentieth Century Architecture*, ed. Ulrich Conrads (London: Lund Humphries, 1970), 168. Hereafter cited as *PM*.

14. Michael Hardt and Antonio Negri, *Multitude: War and Democracy in the Age of Empire* (New York: Penguin, 2004).

Chapter Three

Designations

Mark Jackson

Engaging the work of Michel Foucault, "Designations" approaches the redesign of design through a genealogy that emphasizes the emergence of two discursive fields at the end of the eighteenth century: disciplinary mechanisms for the individuating of selves and apparatuses of security governed by statistical norms. This chapter traces a contemporaneous discourse of aesthetics, engaging the same problem field of inventing relations between individual and population. It locates in the twentieth-century theorist, Walter Benjamin, a concern with bringing these fields into a "constellation," suggesting proximity to aspects of the writings of Martin Heidegger on technology and aesthetics. It also addresses questions of temporality and design broached by both authors.

INTRODUCTION

Signs must become confused where things are entangled. The enslavement of language in prattle is joined by the enslavement of things in folly almost as its inevitable consequence. In this turning away from things, which was enslavement, the plan for the tower of Babel came into being, and linguistic confusion with it (*REA*, 329).

This chapter addresses the work of Walter Benjamin and Michel Foucault. With Benjamin, there is a particular focus on his writing on technology and the artwork in the context of metropolitan cultures, as well as his engagement with language as such and the question of translation. With Foucault, we trace a genealogy of design in the late eighteenth century's concerns with discipline and security that constituted the problematic of "population" as a new object of enquiry and as the object of new technologies of power. There is also the parallel concern with the emergence of aesthetics as an autonomous sphere.

For Benjamin there was something pivotal, something essential to the notion of the *caesura* or interruption, which breaks the continuum of lived experience and flashes the *then* and the *now* in the constellation of a thought image. Benjamin was a thinker of destruction: "construction presupposes destruction."[1] The primordiality of destruction is given some perspective in his 1916 fragment, "On Language as Such and on the Language of Man," concerned with his understanding of the sacred and profane, the melancholy of mute nature and the over-naming of things in the language of man. His engagement with the fall of Adamite man marks Benjamin's writings on modernity and the city. It concerns a fundamental orientation to an understanding of time in the redemptive possibility of experience from the poverty of boredom and distraction. With respect to the question of creativity for Benjamin, we need necessarily to read closely and in conjunction both his "Work of Art in the Age of Mechanical Reproduction," and "The Task of the Translator."[2] While the artwork essay opens directly to an understanding of the city, particularly in the contexts of the cinematic, the translator essay alerts us more carefully to Benjamin's understanding of time with respect to language, meaning and the *caesura*.

With the work of Foucault, the aim is to recognise at the threshold of the nineteenth century the possibility for the emergence of what develops during that century as the discursive field of design.[3] While Foucault addresses aspects of the design of cities, for example, "Designations" engages directly with what Foucault develops as the governmentality of populations. The question of design, and the re-design of design, is approached via Foucault's understanding of governmentality as conditions of possibility for concrete practices of design. Thus part of our enquiry develops an understanding of juridico-legal structures pertaining to territory and things, as disciplinary mechanisms addressing the ordering and confinement of individuals, and apparatuses of security that emerge at the end of the eighteenth century with the invention of "population" as the primary object of governance. Hence our concern with a genealogy of the governmentality of contemporary design as it construes the distributions and dispersions of populations, the ordered mechanisms of exchange and production, and the juridical structures that make coherent and correspondent our artefact world and our identity.

The other preliminary concern is with developing a further understanding of the emergence of the autonomy of aesthetics. This would be a second genealogy, or the second part of the genealogy of modernity's successive approaches to its worlds of appearance in three paradigmatic moments of aesthetic philosophy: those of Immanuel Kant, G. W. F. Hegel, and Martin Heidegger.[4] Clearly this clipped horizon that catches us in a legacy of German Idealism and Romanticism would or could not be the last word. And, in fact, we add Benjamin as the coda to our discussion, and as a way of bringing the two genealogies into a constellation, even as it, perhaps, consolidates further a particularly Germanic perspective. If the discourses of design, from their emergence in the nineteenth century, still grapple with founding design as its own disciplinary domain, they do so from the concrete circumstances of a discursive dispersion across the technological sciences and the fine arts. Any cursory engagement with literature in the field of design thinking over the past twenty years will emphasise the rise and fall of design science as the hopeful pretender to a unification of the field, as well as the curriculum inventions in art history or fine arts institutions of design history and theory programs inveighed by the buoyancy of critical and cultural theory. Hal Foster has documented this well.[5]

Our engagement with the work of Foucault aims to address the extent to which "governmentality" is the bringing into relations of three modalities of power, those of discourses, technologies and practices.[6] We emphasize that concerns with the governmentality of populations as the designed ordering and inventing of technologies of power are non-homologous or non-isomorphic with the fundamental paradigms for the designed appearances of things. It is precisely this legacy as the specific history of our modernity that constitutes the crisis for design's own re-designing. Conservatism in design discourse will find in the non-homology, yet another version of metaphysic's dualism of essence and appearance, or science and aesthetics. We recognise in Benjamin, and in a strained correlation with Heidegger, a significant overcoming of this metaphysical impasse.[7]

TERRITORY, POPULATION, AND SECURITY

In his 1978 lectures at the *Collège de France*, Michel Foucault discussed three discourses and mechanisms of power that have come to define an understanding of the city, or, more generally, the designed designations of the state: juridico-legal structures, disciplinary mechanisms and apparatus of security. The first of these dates from the sixteenth century and we find the emergence of disciplinary mechanisms from juridico-legal structures when the predominant concern of government shifted from the control of territory and things within it to the individuating of populations. This was a shift in the formation and consolidation of cities understood in the emergence of medical and juridical discourses that determined regulations concerned with spatializing segmentations and confinements. By the mid-eighteenth century apparatuses of security emerge as distinct from disciplinary mechanisms when it was recognised that national wealth was not determined by an inventory of the things a nation held, but rather by the productivity of its population. This third mechanism develops as governmentality moved from the spatial problems of confinement associated with disciplinary mechanisms to the temporal problem of planning such that statistics or an inventory of the state increasingly became concerned with forecasting based on the establishment of norms, normativity and normalization as the key concerns of good governance. Crucially, there emerged, with the refinement of the use of statistics, correlations of individuated events that pointed to norms and normativity at aggregated levels of social order. Hence, we see, from the mid-eighteenth century, the "birth" and consolidation of the notion of "population," particularly from the writings of the French economist of the physiocratic school, François Quesnay.

It is not that the discourses of territory and discipline withered in this, far from it. Concerns with territory and population maintained their hold on definitions and discourses of the city in a way that enabled the productivity of relations of power at the level of techniques of normalisation. Foucault stresses the correlative links between the terms "norm," "normativity" and "normalization" with respect to the law, discipline, and security:

> I think it really is necessary to show that the relationship of the law to the norm does in fact indicate that there is something that we could call a normativity intrinsic to any legal imperative, but this normativity intrinsic to the law, perhaps founding the law, cannot be confused with what we are trying to pinpoint here under the name of procedures, processes and techniques of normalization. I would even say instead that, if it is true that the law refers to a norm, and that the role and function of the law therefore—the very operation of the law—is to codify a norm, to carry out a codification in relation to the norm, the problem I am trying to mark out is how techniques of normalisation develop from and below a system of law, in its margins and maybe even against it (*STP*, 56).

A juridical model that refers to a norm is itself produced by mechanisms of normalization, analysed by Foucault across a series of texts, as disciplinary.[8] Discipline is analytical; it breaks down a given situation into minimal elements. It classifies components and establishes optimal sequences. It gives definition to the normal and the abnormal, setting the corrective tasks of the maintenance of normal behaviors and the normatising task of normalizing the abnormal. However, Foucault emphasises the differences between mechanisms of discipline and apparatuses of security precisely at the level of the constitution of the norm itself. If discipline concerns the distinctions and segregations of the normal and abnormal, it is apparatuses of security that constitute the norm itself irreducible to the agency of disciplinary procedures or the individuated subjects thereby produced. With the definitions of "population" a new order of analysis emerges that does not go through the axis of sovereign and subject constitutive of legal structures, nor via the forms of prohibition and control associated with discipline. Rather, what emerges is a form of governance not attached to individuation, but to statistical variance at the level of the social body. It is this variance that will constitute norms:

> The relation between the individual and the collective, between the totality of the social body and its elementary fragments, is made to function in a completely different way; it will function differently in what we call population. The government of population is, I think, completely different from the exercise of sovereignty over the fine grain of individual behaviours. It seems to me that we have two completely different systems of power (*STP*, 66).

GOVERNMENTALITY AND DESIGN

Thus Foucault emphasises that the panoptic mechanism is on the one hand an ideal order for the sovereign, from the earliest of the sovereign states. However, in relation to a system of power that increasingly divorces itself from that of the sovereign, it becomes a new horizon of governance of a population with its specific dimensions and constraints.

This other system of power Foucault terms "governmentality." And it is in the order of governmentality that we may best develop an understanding of the emergence of "design" as discourse, technology and practice of power, but equally as that which engages processes of norm, normativity, and normalization.

"Governmentality" is a concern with design discourse, and technology and practice in the sense that genealogies of the political rationality of governance point to how particular administrative agencies produce their objects of knowledge. The dual objects of governance are "population" and the micro-instrumentality of normalising individuals. These have their correlate in planning the complex relations of individuated elements or sites, infrastructures and limits, to the jurisdiction or definition, for example, of the urban itself. Thus design, or the governmentality of design, is the implicit designating of social goods of the wellbeing of individuated selves and "community."

Foucault's concern is not to explain how the state totalizes in an instrumental and productive way such that governance is secure and complete, but rather how political rationalities of governance, particularly from the late eighteenth century, separate increasingly from the jurisdiction of sovereignty or state power. They cleave themselves from juridical apparatuses, and develop increasingly autonomous technologies of power that define relations between individuals and populations.

Hence his emphasis is on the domain and disciplines of the human and social sciences and discursive fields, which attempt to systematize agencies of normativity with respect to population and individuated selves. It is in this respect that, for example, the disciplines of architecture and planning may be understood as constitutive of political rationalities that are not so much under the jurisdiction of a state, but rather are significant contributors to the complex problematic of the constituency of the state itself. The emergence of a discipline of design does not so much shift the object of analysis, but rather the resolution of the articulation of norms, to the point where, for example, we recognize that there is nothing that is not designed.

DESIGN, LABOR, AND THE UNCANNY

In his *Critique of Aesthetic Judgement*, Kant suggests that the work of art is purposive yet without purpose. By "purposive" he means that it is designed; it is a human design (*CJ*, 61–80). G. W. F. Hegel, in *The Phenomenology of Spirit* emphasizes human emancipation or freedom in the fashioning of things, in human labor, in the shaping of our world (*POS*, 111–19). The work of art is thus an expression of freedom in design or giving shape, in particular, to what is without utility. Martin Heidegger turns to the unworkability of the work, or an essential poetics that does not derive from the intentional will of a human design (*HH*, 123–64).

Kant emphasises the *sensus communis* in an ethics of the artwork as a mimetic design of God's work where the communitarian injunction of an "ought" with respect to the universality of the work of art becomes the very fundament to communitarian projection. We may see the extent to which Kant is working to resolve the difficult relation subtended between individual and population at the end of the eighteenth century precisely by his understanding of the work of art:

> Here I put forward my judgement of taste as an example of the judgement of common sense, and attribute it on that account exemplary validity. Hence common sense is a mere ideal norm. With this as a presupposition, a judgement that accords with it, as well as the delight in an Object expressed in that judgement, is rightly converted into a rule for everyone. For the principle, while it is only subjective, being yet assumed as subjectively universal (a necessary idea for everyone), could, in what concerns the consensus of different judging Subjects, demand universal assent like an objective principle, provided we were assured of our subsumption under it being correct (*CJ*, 84–5).

The state builds its culture on nothing less than this impossible universality, this impossible possibility. Hinged as it is for Kant on there being exemplary works of art that serve as a guide, we recognize the connivance between the institutional frameworks of the metropolitan centers for culture building and the implicit normativity that establishes the state-sanctioned museum as an educative technology of power. Connoisseurship and canonicity established normalizing techniques in appearance constituting the securing modalities of our designed futures.

If for Kant judgement always already resides in the court of reason and hence implies a juridico-legal ground, for Hegel reason is more a motor that moves us. Hence, the slave mentality of the bourgeoisie sublates the non-recognition of the master through labor, which is to say through giving form to unformed matter, and through creating in the stuff of nature the individuated signature of identity. Contra Kant, there is no work or product in the great design of nature, no mimetic faculty of nature perfected. Human labor is the originary font of the work in general, including that of art. We need to read in conjunction two sections of Hegel's *Phenomenology of Spirit*: that concerned with the freedom of self-consciousness and that concerned with the abstract, the living and the spiritual work of art. Hence, in Hegel's discussion of the attainment of self-consciousness in the tale of lordship and bondage, we recognize the importance of labor in freedom. Again, we would also read here Hegel's concern with a crisis in thinking the relation of sovereignty to individual subjects. This implies a thinking of juridico-legal structures in relation to a new concept of power that necessarily constituted itself outside of the circuits of the sovereignty of law. And it concerns nothing less than the artefact production of the world:

> In the lord, the being-for-self is an "other" for the bondsman, or is only *for* him [i.e. is not his own]; in fear, the being-for-self is present in the bondsman himself; in fashioning the thing, he becomes aware that being-for-self belongs to *him*, that he himself exists essentially and actually in his own right. The shape does not become something other than himself through being made external to him; for it is precisely this shape than is his pure being-for-self, which in this externality is seen by him to be the truth. Through this rediscovery of himself by himself, the bondsman realises that it is precisely in his work wherein he seemed to have only an alienated existence that he acquires a mind of his own (*POS*, 118–19).

Hegel stresses the necessity for discipline and servitude for this attainment of the freedom of self-consciousness. We may already anticipate with this quotation the intervention by Marx on the alienation of labor. If the fashioning or design of the artefact constitutes the putting into matter of individuated spirit, such that self-consciousness has its reciprocal recognition, then the fashioning of the work of art, the work without utility or vestige of necessity, will present the possibility of attaining the highest external expression of freedom. This would be superseded by religion as the highest internal expression of freedom; this dialectical relation is further sublated by absolute spirit.

TIME, TECHNOLOGY, AND *POIESIS*

While the artwork will be a passing phase in the restless cunning of Reason, discourses of design and artworks have tenaciously held fast to the authorial functioning of putting spirit to work in matter, and the dominance of the question of form. This is so even, or perhaps particularly, where the artwork, since the mid-twentieth century, has become a question of a time of the work. However, it is Heidegger who has presented a radical engagement with both the Kantian and Hegelian paradigms. Heidegger encounters the artwork as that disclosure of the "worlding" of world, whose materiality is not the matter used, and used up, in the labor of form making, or in the standing reserve of technological production that "challenges-forth" what is, but rather as an "unworkability" of this materiality. This unworkability is not a resistance to the fashioning of usable things, as if the unworkable is a negation of labor. Rather this disclosure radically opens our world of things to us. Heidegger will call it a "bringing-forth," which he recognizes in the pre-Socratic understanding of "*poiesis*."[9]

> Because the essence of technology is nothing technological, essential reflection upon technology and decisive confrontation with it must happen in a realm that is, on the one hand, akin to the essence of technology and, on the other, fundamentally different from it. Such a realm is art. But certainly only if reflection on art, for its part, does not shut its eyes to the constellation of truth after which we are *questioning* (*QCT*, 35).

Heidegger references the German Romantic poet, Friedrich Hölderlin in this essay: "But where danger is, grows / The saving power also" (*QCT*, 34). He is alerting us to the "something" outside of labor as such or outside an anthropocentric and productionist orientation to the question of what is, outside a human will-to-will. In his earlier (1942) discussion of Hölderlin's poem *The Ister*, itself composed at the beginning of the nineteenth century, contemporaneous with Hegel's *Phenomenology of Spirit* and some seventeen years after Kant's *Critique of Judgement*, Heidegger stresses that Hölderlin's poetry stands outside of metaphysics and thus "outside of the essential realm of Western art" (*HH*, 19). Heidegger references Plato and Hegel in order to epitomize Western art's metaphysical foundation:

> With respect to the metaphysical essence of art, we can also say that all art has to do with symbolic images. "Image" ["*Bild*"] then stands for what can be perceived sensuously in general, as can sound. The symbolic "sense" ["*Sinn*"] is the nonsensuous [*das Nichtsinnliche*], which is understood and given meaning and has been determined in manifold ways in the course of metaphysics: the nonsensuous and supersensuous are the spiritual; ideals and "values" are the ideational. The superior and the true are what is sensuously represented in

> the symbolic image. The essence of art stands or falls in accordance with the truth of metaphysics (*HH*, 17–18).

Heidegger's reading will suggest a radical possibility in Hölderlin's poeticising, as a fundamental questioning of being human, or understanding the human as the most "uncanny of beings" the being who arrives at itself only by way of passing through what is most foreign, through translation and the translatability of human being and its being as a whole (*HH*, 48–72). We turn, in fact, to Benjamin for a closer understanding.

LANGUAGE, TRANSLATION, AND THE PROFANE

We have so far outlined a critical program of analysis that would engage with the city as, first, a juridico-legal mechanism at the level of territory and things to be governed; and second as a spatial mechanism of design, where design is predominantly understood as a technology of power concerned with correlates between population and individuation: how for example to segment and make it productive as a positive injunction to happiness. A third arena concerned probability and risk at the level of population and normativity. Thus, how do we foresee what we needed to have done such that we minimize the cost incurred in not knowing? Will to truth as will to power. We have also referenced the emergence of aesthetics at the cusp of the nineteenth century with Kant and Hegel, and Heidegger's reference to this period in the poeticizing of Hölderlin. With Benjamin we recognize immediately an engagement with the question of identity inextricably linked to a question of the metropolitan as an aesthetic experience. Benjamin lodges the artwork not as something in addition to the city that would add to it, nor as something to be missed were cities to abandon a civilizing mission, and not as a task we would have in the progress of emancipation and civility. Rather the artwork is lodged in a primordial engagement of being human in its correspondences with nature that would escape all calculation and labor. For Benjamin there is no autonomous aesthetics. In this, and despite their profound and irreconcilable differences, Benjamin and Heidegger have an uncanny resonance. "The Work of Art in the Age of Mechanical Reproduction" turns to a radical understanding of the city as the political aesthetics of profane illumination brought about by technologies that "temporalise" the immediacy of experience as, for example, an optical unconscious. These same technologies as an aesthetics of politics may equally lead to fascism.

It follows that there are two lines of engagement with Benjamin. One line concerns his understanding of language, developed in a series of essays, but for our purposes, most usefully assayed in his essay "Task of the Translator." The other line of engagement, intersecting with this, is the logic or dialectic of redemption that Benjamin invokes in all of his texts, with respect to the aura, the loss of the aura and redemption, or the sacred, the fall from the sacred, and redemption from that fall in profane illumination. Let us engage with the second one first, as it opens to the primordial concern of Benjamin with language as such. Benjamin's diagnosis of his contemporary culture suggested that something fundamental to experience of one's existence as that which construes cohesiveness in cultural memory is in decay.

Experience becomes something fundamentally partitioned between private, disjointed and non-communicable events, constituting elements of individual repression, and objective existence, inter-subjective but meaningless to an individual. This is the condition of the mass-individual, capitalism and commodity culture, where aesthetics, or concern with the beautiful is also a commodity. Hence for Benjamin there was no point in, for example, differentiating between the artwork and advertising with respect to a former tradition's understanding of the autonomy of art. It is in this condition of decay, of the fall, that Benjamin construes another account of a fall, that of aesthetics itself as it may be mobilized by fascism.

The epitome of the auratic work is the masses as monument to war and death, and the mobilization precisely of photography and film in the aesthetisation of politics. In the profanity of this Benjamin sees redemption, the messianic, the utopian dream, or rather the masses awakening from the dream to the recognition of the dream as dream, to the dream as such. Technicity, reproducibility and the tactile are the profane elements by which the sacred will be redeemed. It is here that I suggest reading between the artwork essay and his essay on the task of the translator is illuminating. The latter essay, in fact, commences with a pivotal comment on the work of art:

> In the appreciation of the work of art or an art form, consideration of the receiver never proves fruitful. Not only is any reference to a certain public or its representatives misleading, but even the concept of an "ideal" receiver is detrimental in the theoretical consideration of art, since all it posits is the existence and nature of man as such. Art, in the same way, posits man's physical and spiritual existence, but in none of its works is it concerned with his response. No poem is intended for the reader, no picture for the beholder, no symphony for the listener (*L*, 69).

In this radical approach to the artwork never intended for reception, Benjamin makes his break with all symbolic functioning as the essence and truth of art's metaphysical heritage. Or, rather, Benjamin will frame the essay on translation along the lines of asking what translation would be in its essence if not primarily for a reader. And, in a repetition of his concerns with origins and reproductions, or with decay and the fall, he will approach the original work and its translation according to a problematic of profane illumination. The authentic relation between an original and its translation is that of essential correspondences between languages that reveal, not the transmission of meanings from language to language, but pure language, language as such. Fidelity in translation comports itself to this essence.

The translator text serves as an introduction to a translation of Baudelaire's *Tableaux Parisiens*. Benjamin's late writing on Baudelaire, "On Some Motifs in Baudelaire," re-inaugurates a notion of the aura precisely in the hidden correspondences it reveals such that the play essential for redemption has its potentials.[10] He thus suggests that the aura's experience lies in the correspondence of the inanimate or nature to men. Such correspondences are unconscious relations of the human to nature, explored for example by Benjamin in his early essays on language as a mimetic faculty.[11] Thus Benjamin emphasises that there must be a human element in things, which is not brought about by labour, an understanding uncannily approximate to Heidegger when approaching the essence of technology and the artwork. Both extol or privilege Hölderlin, at times in almost interchangeable language, concerning translation, identity, and the foreign. If Benjamin's artwork essay was concerned with a certain loss of aura, this aura concerned an essential spatiality of the near and the far, of proximity and place.

Profane redemption from this loss happens in the medium of a temporal work. In cinema there is the "dynamite of the tenth of a second" blows apart what was previously contained and locked in (*I*, 238). This redemption reveals a politics of time. If the "Task of the Translator" suggests a new aura, explored in Baudelaire's Paris, it is not a return to the near and far of a spatiality of aura, or to confinements, disciplinary spaces and segmentations of populations, but rather to a temporality of being human revealed in the mimetic faculty of an optical unconscious, in Benjamin's terms, a lightning-flash of recognition, or an ecstatic temporality in Heidegger's.

Such a politics of time engages a question of the de-signing of design, in the designation of design, not from the viewpoint of a form giving or trace-inscription, writing in its most general sense. Rather, such a politics of time engages with the "now-time" of Benjamin's shock of recognition, the profane redemption of what has been in the light of a messianic future. If we infer such an engagement offers radicality, it is not because it has fundamentally displaced an auratic spatiality by an auratic temporality. Foucault has already made clear that in as much as security liberates the contained flows of disciplinary mechanisms, it is fundamentally a temporality of a projected future that founds the precepts of liberalism.

Crucial to a political aesthetics opened by Benjamin, correlative to Heidegger's reading of translation and the foreign in *The Ister*, is that its target is precisely those regimes of truth established by what Foucault has identified as emergent from the late eighteenth century as apparatuses of security. The temporality of Benjamin's "now-time" is that of a *caesura* or halt to the secured risk-assessment of our future, which is to say the making of our future into a permanent present of evaluation and determination, what Heidegger might call the set up or en-framing of a standing reserve (*QCT*, 19–20). Benjaminian political aesthetics would construe the singularity of a mass individual, with its tendency to fascism, into multiple sites of tactical resistances to the technologies of security constitutive of the normalizing programs of design. De-signing would then constitute not a series of acts, potential or actualized, but rather something outside the field or agency of a laboring self, as a receptive openness to the mimetic correspondences of the human to the inorganic that is the belonging of our being no longer to a will to will.

NOTES

1. See, for example, Benjamin, "The Destructive Character," in *Reflections: Essays, Aphorisms, Autobiographical Writings*, ed. Peter Demetz (New York: Schocken Books, 1986), 301–3.

2. Walter Benjamin, "The Task of the Translator," in *Illuminations*, ed. Hannah Arendt, trans. Harry Zohn (London: Fontana, 1973), 69–82; and "The Work of Art in the Age of Mechanical Reproduction," in *Illuminations*, ed. Hannah Arendt, trans. Harry Zohn (London: Fontana, 1973), 219–54. Hereafter cited as *I*.

3. There is a particular focus on Foucault's lectures from the College de France. See, for example, Michel Foucault, *Security, Territory, Population: Lectures at the Collège de France, 1977–1978*, ed. Michel Senellart, trans. Graham Burchell (Hampshire and New York: Palgrave Macmillan, 2007). Hereafter cited as *STP*.

4. See Immanuel Kant, *The Critique of Judgement*, trans. James Creed Meredith (Oxford: Clarendon Press, 1952). Hereafter cited as *CJ*; G. W. F. Hegel, *Phenomenology of Spirit*, trans. A. V. Miller (Oxford: Oxford University Press, 1977). Hereafter cited as *POS*; Martin Heidegger, *Hölderlin's Hymn* "The Ister," trans. William McNeill and Julia Davis (Bloomington Indianapolis: Indiana University Press, 1996). Hereafter cited as *HH*. Heidegger's 1942 lecture course that reads Hölderlin's poem presents one of the most radical engagements with the questioning of the work of art, an engagement that significantly revises Heidegger's earlier treatment of Hölderlin in his 1935 *What is Metaphysics* and *The Origin of the W ork of Art* versions, which were also written in the early to mid-1930s.

5. See Hal Foster, "The Archive without Museums," in *October* 77 (Summer 1996): 97–119. Also his book length study *Design and Crime (and Other Diatribes)* (London/New York: Verso, 2002). With respect to design theory and texts on the discipline of design, see, for example, Bruce Archer, "A View of the Nature of Design Research," in *The Proceedings of the Design Research Society*, ed. Jacques R. and J. Powell (Portsmouth: Design Research Society, 1981): 30–47; Victor Margolin, *The Politics of the Artificial* (Chicago: University of Chicago Press, 2002); Nigel Cross, *Designerly Ways of Knowing* (London: Springer Verlag, 2006); Klaus Krippendorff, *The Semantic Turn: A New Foundation for Design* (Boca Raton, FL: Taylor & Francis, 2006).

6. See the series of essays in *The Foucault Effect: Studies in Governmentality*, ed. Graham Burchell, Colin Gordon and Peter Miller (Chicago: University of Chicago Press, 1991). Hereafter cited as *FES*. See in particular, Michel Foucault, "Governmentality," *FES*, 87–104; and Colin Gordon, "Governmental Rationality," *FES*, 1–52.

7. This "strained correlation" refers to our reading across Benjamin's "Task of the Translator" and Heidegger's *The Ister* for a concern with the work of art that no longer goes via the circuit of labor, production, or anthropocentrism for that matter. This radical opening across the incongruity of Benjamin and Heidegger suggests an approach to design that would itself be non-reducible to technological aesthetics. The common resonance rests with a radical understanding of translation and translatability.

8. See, for example, Foucault's two important studies on practices of confinement: *The Birth of the Clinic*, trans. Alan Sheridan (London: Tavistock, 1973), and *Discipline and Punish: The Birth of the Prison*, trans. Alan Sheridan (London: Allen Lane, 1977).

9. Martin Heidegger, "The Question Concerning Technology," in *The Question Concerning Technology and Other Essays*, trans. William Lovitt (New York: Harper Torchbooks, 1977), 3–35. Hereafter cited as *QCT*. On Heidegger's understanding of the relation of the art work to equipment and the essence of equipmentality in reliability, see "The Origin of the Work of Art," in *Basic Writings*, ed. David Farrell Krell (London: Routledge, 1993), 139–212.

10. Walter Benjamin, "On Some Motifs in Baudelaire," in *Illuminations*, ed. HannahArendt, trans. Harry Zohn (London: Fontana, 1973), 157–202.

11. Walter Benjamin, "On the Mimetic Faculty," in *Reflections: Essays, Aphorisms, Autobiographical Writings*, ed. Peter Demetz (New York: Schocken Books, 1986), 333–36.

Chapter Four

Signs of Postmemory in Dresden

Restoring the Displaced

Marsha Berry

This chapter narrates my encounters with a past accessible only through the stories and images of significant others, as I recount my wanderings through Dresden, a place of postmemory for me as a child of exiles (known as "displaced persons") who emigrated to Australia in 1950. I draw on the work of memory studies scholars such as Marianne Hirsch, Aleida Assamann, and Andreas Huyssen to locate my experiences. The discourses of postmemory strongly influence how we encounter places that are overlaid with traces of monumental historical events such as World War Two. I read Dresden against these discourses to conceptualise relationships between postmemory aesthetics, places and imagined local knowledge and how these relationships may be used for restorative creative practice.

INTRODUCTION

> I know that my parents will soon be back from their holiday, and there is something important which I must give them. I am not aware that they have been dead for years. . . . But when at last they come through the door they are in their mid-thirties at the most. They enter the flat, walk around the rooms picking up this and that, sit in the drawing room for a while and talk in the mysterious language of deaf-mutes. They take no notice of me. . . . It does not seem to me, Austerlitz added that we understand the laws governing the return of the past . . . As far back as I can remember, said Austerlitz, I have always felt as if I had no place in reality. . . . (*A*, 184–85).

I had attended a conference in Leipzig and was driving on the autobahn with my cousin on the way to visit family in Dresden feeling overwhelmed by nostalgia and a sense of déjà vu. It was my first time in the former German Democratic Republic (East Germany). Dresden is a city that evokes a palimpsest of a glorious, Baroque, architectural past, over which are inscribed black and white images of the aftermath of firebombing in the final months of World War Two. The city was also behind the Iron Curtain until the 1989 fall of the Berlin Wall. It is a city that brings to mind "spatial imaginaries" inhabited by ghostly images, a city haunted by its traumatic past (*PPU*, 10). My arrival in Dresden in 2007 felt like a return, yet I had never been there. I was experiencing the collision of memories: an overlay of inherited personal memory and public memory. How was this happening?

As a child of exiles (displaced persons), my mother told me many stories about Germany, especially about Berlin where she grew up, and Dresden, from which she miraculously escaped in February 1945. She never saw Dresden again as it was in the Soviet Bloc. The stories of Dresden were an integral part of my childhood. I had clear visual images of Dresden as projected memories and geographies. There were gaps as well as a melancholic sense that I could never share in such a place of beauty because I was born too late, out of time, my mother's time. I was the seed for a new country, a new place, new home.

Marianne Hirsch coined the term postmemory to theorize the experiences of the children of survivors of mass trauma such as the Holocaust. Such children grow up with narratives of traumatic events that happened before they were born. The narratives become ingrained in early childhood as a pervasive and seductive discourse, a background against which lived experience becomes alive and is interpreted. Memories become imagined and projected onto places from which survivors fled or from which they were exiled. Photographs and film footage are key characteristics of the artworks and installations Hirsch uses to expose the discourses of postmemory.

My mother's photographic albums are filled with scenes of forests and lakes with family, friends, and acquaintances of her youth in Germany during World War Two. Her father was a photographer and she too had a camera, a Leica, so she had many black and white photographs; there was a story attached to each one. Some of the people in the photographs had died young in the war, or were overtaken by tuberculosis, some had immigrated to America, some remained in Europe, and others simply disappeared from her life. These photographs became significant threads in the texture of my childhood influencing the contours of my projected memories that at times overshadowed memories of direct experiences.

Postmemory as a phenomenon is associated with place in that the discourses of postmemory strongly influence how we encounter places that are inlaid with traces of monumental historical events such as World War Two.

Vicarious local knowledge as an imagined construct filters real time encounters displacing the here and now with the way things were or might have been in another time. Such imaged local knowledge of place built by the children of survivors through stories and photographs are replete with gaps. These absences add layers to the ways we encounter place and to what constitutes local knowledge.

Postmemory may be described as a form of memory that is powerful "because its connection to its object or source is mediated not through recollection but through imaginative investment and creation" (*PT*, 662). It is associated with exile and diaspora, and often with family photographs. Hirsch distinguishes two types of Holocaust photographs: those depicting Holocaust events such as mass graves, and those depicting family celebrations and portraits. She argues that we view the latter with a sense of disbelief once we know the context, particularly when we realise that the people in them became faceless victims of the Holocaust. Many of the people pictured in European, Jewish, family photographs taken in the 1930s and 1940s perished in Hitler's death camps. These images are also critical to postmemory because they recreate that which has been lost through the exile and diaspora that follows mass trauma. Hirsch supports her contention that family photographs with domestic themes from the time of the Holocaust may be regarded as Holocaust photographs even though their subject matter is not directly concerned with the atrocities. Hirsch describes her experience of viewing "Tower of Faces" at the United States Holocaust Memorial Museum, specifically her viewing of the Lithuanian *shtetl* of Ejszyzski photographs that depict ordinary people marking "life's transitions" (*PT*, 669): the contents of Ejszyzski photographs are not explicitly about the horrors of the Holocaust; rather they evoke ordinary life as it was prior to the Holocaust. She notes that the photographs evoke empathy as children of survivors start to identify with the figures in the photographs so that the "Tower of Faces" tells their family's story as well, by proxy. The experience had a profound impact on Hirsch.

Photographs as visual signs then are critical to the formation of postmemory. In *On Photography* Susan Sontag describes her encounters as a child with her father's photographs of World War Two and Nazi atrocities. Sontag's view of the world was transformed by the images: it was a profound transformation. We do not have to see the images to know the content. Not only do the images have a significant emotional impact on survivors of such mass traumas, but they also have a profound influence on the children of survivors who may be settled quite comfortably in a new country. As a starting point for building a theory of postmemory, Hirsch argues that the shock of the images has a different effect on the second generation (*PT*, 669). She defines postmemory as temporally and spatially different from survivor memory. It has characteristics of displacement, belatedness, projection, and vicariousness, and is mediated not through direct remembering but through imagination and empathy. Postmemory creates a different set of relations with past experience whereby the experience of another's experience becomes an experience in itself. Places overlaid with traces of survivor memories and images are central to postmemory. In this way both experiences and places are a part of the second generation's consciousness with experiences of placement and displacement. And photographs that help constitute postmemory do not need to have traumatic contents; they often operate as palimpsests.

Postmemory is a dominant conceptual theme in much recent art and photography in Europe. Ulrich Baer interrogates Rienatz's photography arriving at the conclusion that his photograph of a clearing in a forest, *Sobitor Extermination Camp Grounds* "restores a sense of place" to Holocaust events that appear "geographically and historically placeless" (*GR*, 46). It is difficult to view the photograph without reading it against postmemory. The work relies on survivor memories and postmemory discourses surrounding Eastern European forests for it holds an ambience of arbitrary destruction.

The aesthetics of postmemory rely on acts of memory and imagination. For example, the artist Christian Boltanski seeks to invoke a postmemory aesthetic in his powerful installations. His aesthetic and creative practices perform acts of remembering expressive of his knowing and conflicted feelings and experiences. In a 1995 interview Boltanski states, "my work is bound to a certain world that is bordered by the White Sea and the Black, a mythic world that doesn't exist, that I never knew, a sort of great plain where armies clashed and where Jews of my culture lived" (*PT*, 679). The words and works reveal tensions created in those who have been born after happenstances that led to exile and diaspora. His postmemory is associated with place and a sense of displacement. He interrogates the gaps between postmemory and memory, both individual and collective. Through his art practice he seeks to install exile and dislocation as a form of collective memory that features emptiness as a dominant aesthetic.

Postmemory discourses and aesthetics are concerned predominantly with the Holocaust; however, others have departed from the Holocaust to explore the impact of photographic imagery of other monumental traumas, such as World War One[1] and the American War in Vietnam.[2] Clearly Hirsch's focus is on the Holocaust, however she herself states that she does not "want to restrict the notion of postmemory to the remembrance of the Holocaust, or to privilege the Holocaust as unique or limit experience beyond all others" (*YJC*, 11). Postmemory is therefore about looking back, adopting and exposing collective traumatic experiences pertaining to a culture. Dresden is a place of monumental trauma and there are iconic images showing the devastation of the fire bombings inscribed in collective memories of World War Two. In commencing my reading of Dresden, it is as a postmemory place with memory of my family's memories of suffering: my "retrospective witnessing by adoption" (*YJC*, 10).

POSTMEMORY FEBRUARY 13, 1945

The point of origin for my reading of Dresden is a temporal line stretching between February 1945 and March 2007. My postmemory of Dresden became grounded in March 2007 when I was able to visit the former East Germany. My parents immigrated to Australia in 1950 as displaced persons from a German refugee camp. Both were former Soviet citizens escaping from Stalinist policies. My mother was the sixth child in a family of seven. In 1929 her father was incarcerated by the Soviets under Sections 58.10 to 58.11 for counter-revolutionary activities and sent to the Solovetsky Islands (Solovki) near Karelia to the gulags. In 1933 he was released. The family went to Berlin to live with my grandmother's parents—exiled. My mother felt she had escaped one bad situation to land in another, that of Hitler's Germany. Her war was a tale of travel and epic stories of escape.

As a small child living in Melbourne's Surrey Hills, I heard many conversations my mother had with my father and aunt reminiscing about their wartime experiences. The stories igniting my imagination were those of the fires that often came after bombings. One night my mother ran from apartment house to apartment house trying to find somewhere safe that was not alight. This recollection became part of my memories igniting recurring childhood nightmares of European streetscapes where fires raged, leaping from house to house. My sister, charged with babysitting, would drag me under our Surrey Hills' dining room or kitchen table whenever an aeroplane went overheard serving to exacerbate those fearful memories.

In 1944 my mother was living in Dresden. I remember her narration of inner conflict about leaving Dresden the morning of the night of the first bombings. I also remember her unspoken emotion. My sister was two years old in 1945. To leave they needed to cross a front, and my mother would need to carry my sister on her back as well as a rucksack with clothes and essential belongings. The Soviets were close. The rumours of how the Red Army treated civilians were horrifying. As an exile from the Soviet Union she needed to flee westwards toward the Allied forces. But the apartment was comfortable and she was tired of moving. Her sense of impending disaster prevailed. She left and walked all day. At nightfall she found a ditch in which to bivouac for the night. She watched the multitude of planes flying overhead toward Dresden and then the orange glow on the horizon. Dresden was burning. The awful beauty burned into her memory. She was fully aware of what residents were facing in both the firestorms and the advancing Red Army. Dresden would be in the Eastern Bloc. She would never go back. She knew that night as she watched the firestorms that she had irrevocably lost her home and her sense of being at home in Germany.

Eyewitness accounts provide testimony to the horror of the firestorms witnessed by my mother from a relatively safe distance. Lothar Metzer describes his memory to Tim Halloway in a recorded eyewitness account:

> It is not possible to describe! Explosion after explosion. It was beyond belief, worse than the blackest nightmare. So many people were horribly burnt and injured. It became more and more difficult to breathe. It was dark and all of us tried to leave this cellar with inconceivable panic. Dead and dying people were trampled upon, luggage was left or snatched up out of our hands by rescuers. The basket with our twins covered with wet cloths was snatched up out of my mother's hands and we were pushed upstairs by the people behind us. We saw the burning street, the falling ruins and the terrible firestorm.[3]

The bombing of Dresden has attained monumental status in the collective memory and its import was felt soon after it occurred. Photographs taken in the aftermath of the firebombing serve an emblematic function. Images of figures examining the corpses in the streets of Dresden engage the viewer's imagination and empathy leading one to wonder what it must have been like to pick over corpses looking for a loved one and the mixed emotions of revulsion, despair, and hope. Aerial shots of bombed-out ruins of once beautiful buildings, as well as churches and public squares evoke the scale of destruction and loss of life. One is witnessing the horror of human destruction and the injustices perpetrated on innocent citizens going about their daily lives.

In the West a major debate followed the bombing of Dresden. Winston Churchill attempted to distance himself from the bombings by writing the following in a memorandum to the British Chiefs of Staff:

> It seems to me that the moment has come when the question of bombing the German cities simply for the sake of increasing the terror, though under other pretexts, should be reviewed. . . . The destruction of Dresden remains a serious query against the conduct of the Allied bombing.[4]

The traumatic significance of the bombings remains. British historian, Frederick Taylor, in an interview with *Speigel Online*, likens the bombing of Dresden to an epic tragedy:

> The destruction of Dresden has an epically tragic quality to it. It was a wonderfully beautiful city and a symbol of baroque humanism and all that was best in Germany. It also contained all of the worst from Germany during the Nazi period. In that sense it is an absolutely exemplary tragedy for the horrors of 20th century warfare and a symbol of destruction.[5]

The fire bombings of cities such as Dresden and subsequent invasion by the Soviet army have recently become a dominant discourse for art, films, and popular media in Germany. Stories of these traumas have been told continuously within families, narratives creating postmemories for many Germans. This process has been further complicated by the post-War existence of East and West Germany. Memory scholar Aleida Assmann analyzes the current shifts in German memory concerning the suffering at the end of World War Two. She starts with the proposition that historical traumas may often reappear belatedly and that this is because of "the trauma's resistance to representation, which involves mental blocking and psychic dissociation, as well as social and political taboos" (*GLL*, 187). Assmann observes that many second and third generation writers and artists in Germany are engaging with the politics of postmemory discourses. Many are meeting the phenomena of belated postmemory in Germany with anxiety fearing that a focus on the suffering of the German people at the hands of the Allies may alleviate their collective guilt. Assmann examines this anxiety and refers to Nuremberg Trial court records to show how the logic may be perverted to position perpetrators as victims thereby expunging their guilt. She proposes:

> In any society highly divergent memories and group experiences exist always side by side which do not create conflicts because they are not elevated to a public level; on the level of public discourse and national identity, however, the question arises as to how one can integrate divergent and even contradictory memories into a generally acceptable framework. In Germany the solution to the problem of the heterogeneity of memories is, I believe, to be found in their hierarchical struggle. . . . Memories exist and are constructed on individual, family, social and national levels (*GLL*, 197).

She concludes that the dynamics of memory cannot be stopped through discourse rules and taboo. I would suggest that memories could give rise to postmemories that call for expression. Memories should not be repressed, but rather be expressed and contested. Through the challenges of reading problematic and painful places like Dresden, via postmemory discourses, contested memories may find utterance. Suffering does not cancel nor excuse the guilt of Hitler's Germany; nevertheless Soviet soldiers raped the women in Dresden, and women and children were vaporized by fire.

I agree with the sentiments of contemporary British philosopher, Anthony Grayling, that the bombing of Dresden was horribly traumatic because "destroying everything . . . contravenes every moral and humanitarian principle debated with the just conduct of war."[6] The discussion of what constitutes a just war will not be addressed here and indeed is not specifically relevant to a discussion of postmemory discourses and how they frame experience of place. Rather my focus is on how places contain ghostly traces of traumatic events and how these can inspire creative practices. I was inspired by my mother's timely escape from Dresden.

PLACEMENT/DISPLACEMENT

The photographic images of my postmemory were black and white. They were imbued with a sense of nostalgia as well as nightmare. Leo Spitzer describes family photographs taken at the Austrian Club in La Paz in Bolivia in 1947.[7] He perceives their irony in that the people depicted revel in their Austrian-ness even though they were refugees who had fled Austria after the *Anschluss* and Nazi persecution. He interrogates nostalgic memory in order to interpret the images. Nostalgia or *Heinweh* (literally translated "home ache") is a crucial ingredient of postmemory. Nostalgia, according to Spitzer, has a therapeutic quality for refugees allowing them to construct narratives about the past in order to adjust to life in the new country and ease those feelings of culture shock and alienation. French sociologist Maurice Halbwach also noted this phenomenon, claiming that nostalgia is able to liberate people from the constraints of time so they can stress positive aspects of the past through selective remembering.[8] Nostalgia is not merely a pleasant melancholic longing for a past; it is a reconstruction of memories and associated narratives that can assist with the integration of monumental traumatic events in order to adjust to exile.

Photographs were an important part of my family's memorabilia triggering both the nostalgia and relief of refugees. Germany was always "back home," but it was also "back then before we came here and you were born." And the colors of nostalgia for me were always black and white. I had been to Germany before, but not to where my mother had lived. My lived experience of Dresden in March 2007 was haunted by my mother's escape. I projected feelings of displacement and connection with place and projected local knowledge from my mother as well as iconic images of bombings and a demolished Frauenkirche. I expected Dresden to be a sad monochrome place.

Instead, I encountered a city of color and restoration. This was a shock acting as a device of estrangement: the familiar looked unfamiliar because it was in color. Frauenkirche had emerged from the ashes like a phoenix. The city appeared as a sensuous environment with public places and spaces that seemed to encourage the processions of bodies to stop, to look, and engage with something other than daily banalities. Places to sit were positioned strategically to encourage people to stay and deepen their sense of history, and to form new memories of place. I began to have a strong impression that it was as though the Dresden of pre-February 13, 1945, indeed the pre-Nazi Dresden had a melancholy voice that wanted to overlay these events with hope for the future grounded in its Protestant Baroque past. For me this voice found expression in the public square surrounding Frauenkirche. Voices of postmemory began to compete with older voices from the epochs of nineteenth-century Impressionism and seventeenth-century Saxon Baroque. Figures dressed in Saxon Baroque costumes paraded outside Frauenkirche handing leaflets about the church to passing tourists. For a moment Baroque voices were the loudest.

I began a visual record comprising video and photographs of my experience of Dresden as a public space. Connecting my postmemory and memory of this place, these will always serve as souvenirs. The blackened stones, visible in the photographs, remain as a deliberate memorial to the Frauenkirche prior to the fire bombings. But memorials can become invisible and forgotten through habit. As Huyssen observes in his discussion of the crisis of memory and memorials, "the act of remembering is always in and of the present, while its referent is of the past and thus absent. Inevitably, every act of memory carries with it a dimension of betrayal, forgetting, and absence" (*PPU*, 3–4). I wondered how long it would be until the blackened stones vanished, forgotten in the urban palimpsest that is Dresden. For me the stones would remain visible and present as referents to the violent history of the twentieth century.

According to Roland Barthes, "the photograph is literally an emanation of the referent. From a real body, which was there, proceed radiations which ultimately touch me, who am here" (*CL*, 80). I had been touched by the photographs of Dresden after the fire bombings and now I was touched by a material presence restoring that which once was, was no more and is again. What I was looking at, touching, photographing was a restoration moving me to tears of gladness, and at the same time a belated sense of grief and anger at the actions of the Nazis who had, effectively, robbed my parents of their youth and promoted their status as displaced persons. My encounter with the material presence of Frauenkirche allowed me to forgive those who had caused my mother's suffering and displacement—with a sense of restoration on many levels.

I wandered through the old city of Dresden with my cousin, Sacha. Interestingly we heard many tourists speak Russian, Polish, and Czech, languages of the old Soviet Bloc. I could not shake the sense of familiarity—Dresden after all follows the pattern of many European cities built around a river with a majestic center comprising churches, palaces and public buildings. I found I could "read" the city much to the surprise of my cousin who had grown up in Belgium and thought Australia was at the ends of the earth with cities that were completely alien to the European notion of a city. Around each corner was a sight more magical than the last. We abandoned maps and walked as our fancy took us, stopping regularly for coffee and cake. We gloried in the beauty and acknowledged the labors of those who had gone before to create a built environment on a majestic yet human scale. We reminisced about the stories our mothers, who were sisters, had told us about World War Two and we drank in the sights of Stallhof and Schlossplatz. We sat in the Royal Mews listening to a busker playing a violin—coincidently playing some Russian tunes including *Moscow Nights*: for us, a profound moment of synchronicity. We felt my mother's presence and missed the possibility to ask her about the gaps in our postmemories and what she thought of the restorations. It was too late. My mother had passed away in 1998. In that moment I realized that Dresden was no longer simply a place of imaginings for me. My experience of place and space was my own, informed by postmemory, but no longer constituted by it. Dresden was no longer a place distant in time extant only through my projections and my mother's memories; it was now truly a part of my lived experience and future narratives. I felt the sorrow of displacement anew, not as a projection but rather as an actuality: no longer on my mother's behalf but on my own. Her story has ended and mine is still a work in progress.

Many traces of the firestorms of February 1945 remain in the form of burnt-out mansions, blackened churches, and palaces. They enhanced our sense of belatedness and an overwhelming abundance of life and death stories. The neglected hulks of burnt-out houses, apparently untouched for sixty years are incredibly sad, yet have a heroic and almost stoical presence. Later we found out from my nephew, an architect in Dresden, that many such dwellings are left until rightful owners or their descendants can be found. The original owners abandoned them to flee to British and American sectors in 1945: an echo of my mother's flight on the day of February 13, 1945. They were left as reminders in the form of propaganda by the Soviets. Now they stand as architectural orphans waiting to be claimed so that, once again, they can fulfil their function as residential spaces.

Cycles turn. My family remains connected with Dresden; my nephew lives there with his wife. They specialise in restoration, making contemporary living spaces in old apartment houses. The church frequented by my mother remains, albeit surrounded by utilitarian Soviet architecture known as *Plattenbauten.*[9] Nonetheless, the church with its cheerful blue domes, built in 1872–1874 for the Russian legation to the Kingdom of Saxony, still has room to breathe; it still keeps its place in this city of epic postmemories. We stepped inside and lit candles to the memories of those no longer present, but whose memories had guided us to this place.

My mother's feelings about the bombing of Dresden were conflicted. She was an alien in Nazi Germany and inwardly cheered each time the Allies bombed German cities, even when she sat huddled in basement bomb shelters. She never felt a victim of Allied bombings, rather she felt they were a necessary evil to be endured, and hopefully survived, in order to free Europe of Nazi oppression. Yet she mourned the destruction of Dresden's heart and the vision so eloquently portrayed by Canaletto paintings and captured in early twentieth-century photographs.

PLENITUDE AND GAPS

Hirsch proposes that children of exile suffer from deep feelings of displacement concerning memories of monumental places to which there is no going back and this in turn generates a sense of plenitude rather than absence (*PT*, 11). The sense of places being too full of stories, feelings, and memories creates gaps and fissures that add to their texture. She raises questions about the aesthetic shapes in Holocaust art influenced by postmemory discourses. I too, felt overwhelmed by both plenitude and gaps in the texture of Dresden and sought an aesthetic shape in the form of video art to express this response. My raw material comprised footage and shots of places in Dresden using my vicarious, inadequate local knowledge; and, an old home movie showing my mother as a twenty-year-old woman (a woman I knew and never met save in photographs and my imagination as she was thirty-four when I was born).

The *Displaced One* uses a point of view that leaves gaps for the viewer through editing together the old and new. The work is a montage of decaying colour film footage shot in Berlin in 1940 and subsequently digitized, and juxtaposed with mobile phone footage I shot in Dresden. The piece opens to show a man walking backward in slow motion up a set of steps to a door. I used backward motion to reference the impossibility of returning to a point before monumental events occur. A woman (my mother) joins the man and they walk away together before parting company. During the War people would part ways, always uncertain if they would see each other again. This I highlighted with a dissolve to a pan of the Royal Mews in Dresden; it did not provide an obvious narrator point of view. The narrator is my postmemory. The viewer does not see the woman again; rather the viewer is presented with an absence, the nostalgic absence felt by the second generation. The soundtrack is an accompanied male voice singing *Moscow Nights*: a reference to the exile of my mother, her siblings and parents and homesickness for a place in Russia that no longer existed.

The questions of what happened to the man and the woman are never answered. This echoes the experience of many children of survivors as they looked through family albums at people who were part of the family circle before the mass trauma; and always the question with no answer, "What happened to so and so? I wonder what became of him/her." The square is the central "character" in this scene. The gaps in between events are left to the imagination of the viewer. This is what a postmemory aesthetic may look like with the narrative in a state of beginning, yet becoming absent in a place that still remains intact. Dresden is still there. But the space in Dresden inhabited by my mother is long past. There is no return.

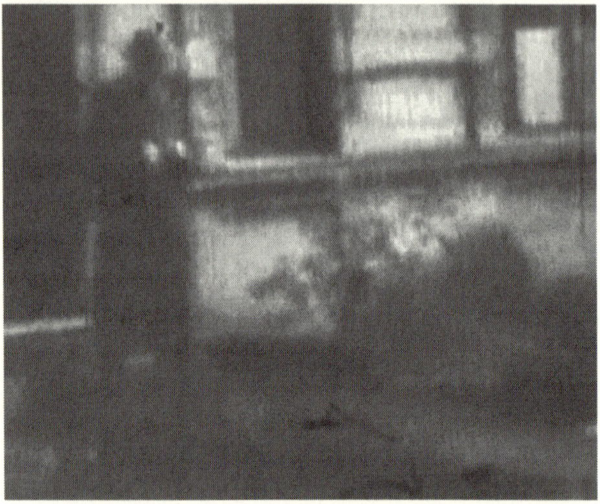

Figure 4.1. Restored film from *Displaced One*, Marsha Berry, 2008.

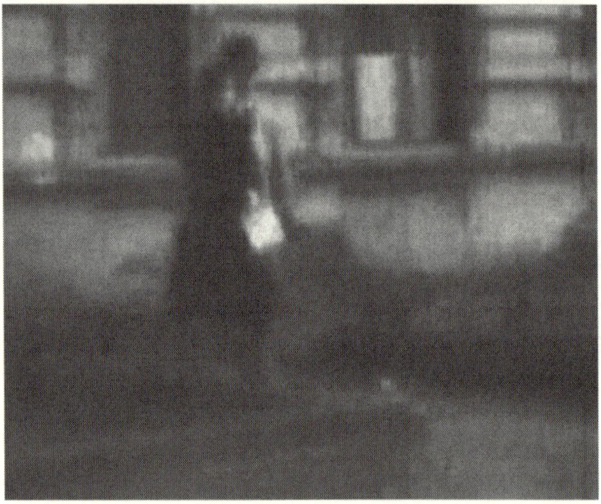

Figure 4.2. Restored film from *Displaced One*, Marsha Berry, 2008.

Figure 4.3. Cross-dissolve between restored film and new mobile phone movie from *Displaced One*, Marsha Berry, 2008.

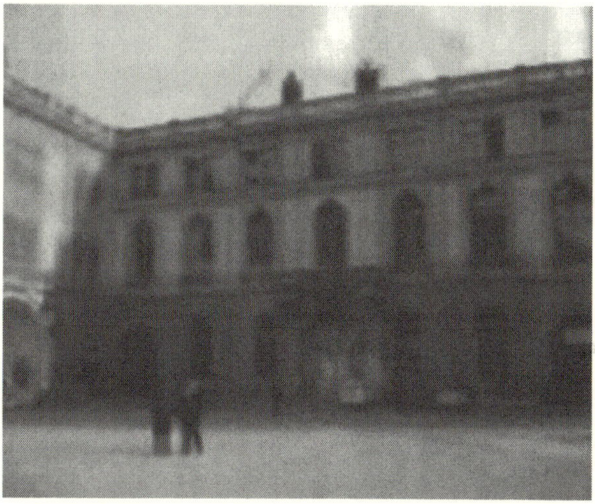

Figure 4.4. New mobile phone movie from *Displaced One*, Marsha Berry, 2008.

NOTES

1. Marlene A. Briggs, "The Return of the Aura: Contemporary Writers Look Back at the First World War Photograph," in *Locating Memory*, ed. Annette Kuhn and Kirsten Emiko McAllister (New York: Berghahn Books, 2006), 113–34.

2. Patrick Hagopian, "Vietnam War Photography as a Locus of Memory," in *Locating Memory*, ed. Annette Kuhn and Kirsten Emiko McAllister (New York: Berghahn Books, 2006), 201–22.

3. Tim Halloway, “The Fire-bombing of Dresden: An eyewitness account,” *Timewitnesses*, recorded May 2009, accessed November 29, 2007, http://timewitnesses.org/english/~lothar.html.

4. Detlef Seibert, “British Bombing Strategy in World War Two,” last modified February 17, 2001, accessed January 8, 2008, http://www.bbc.co.uk/history/worldwars/wwtwo/area_bombing_01.shtml.

5. Charles Hawley, “Dresden Bombing to be Regretted Enormously,” interview with Frederick Taylor, *Spiegel Online*, last modified February 11, 2005, accessed November 29, 2007, http://www.spiegel.de/international/0,1518,341239,00.html.

6. Anthony Clifford Grayling, *Among the Dead Cities: Was the Allied Bombing of Civilians in WWII a Necessity or a Crime?* (London: Bloomsbury, 2006).

7. Leo Spitzer, “Back Through the Future: Nostalgic Memory and Critical Memory in a Refuge from Nazism,” in *Acts of Memory*, ed. Mieke Bal et al. (Hanover: University Press of New England, 1999), 87–104.

8. Maurice Halbwachs, *Les Cadres Sociaux de la Memoire* (Paris: Alcan, 1925).

9. Nicholas, Howe, “Kilroy in Dresden,” *Dissent* (Spring 2001), http://www.dissentmagazine.org/article/?article=1267.

Chapter Five

Posed Solitude

Signing a Poetics of Community

Maria O'Connor

This chapter approaches the thinking of Maurice Blanchot's *récit*, Martin Heidegger's *ecstatic temporality* and Marc Augé's *non-place* in the questioning of Blanchot's *unavowable community* as a site of politico-ethical poetics conditioned by his understanding of work (*oeuvre*) and worklessness (*dé-soeuvrement*). This poetics as community engages the posed solitude of Augé's non-place. Augé's posed-solitude is read here as a more ecstatic relation between self and other beyond conditions of dialecticity in a temporality that relates to Heidegger. The *neutral* as a spatiality encountered in Blanchot's writing and community and the radicality of Augé's non- place open an horizon for spatial design practices. Such practices recognise the politico-ethical responsibilities for the legacies of modernity's failed communities of identity and the most immediate political imperative of unavowable communities: communities of citizens without citizenship, refugees, forced migrants, dispossessed, and exiled peoples. Design, poetics, and community coalesce here, before failure that cannot be collapsed easily into frameworks of morality and judgement.

INTRODUCTION

What is it to say that community is poetics? Where or how does this community exist? Perhaps, there is a commonly conceived notion of community as that entity or network that activates spaces for conditions of familiarity. This would be a community that is determinable and known, a community that is in some way pre-planned into the fabric of existing urban spaces, designing an overarching facilitation of identities that come together. For example, in the conception of today's urban planning, mastery is desired in the design to reconstitute urban sites that are productive of communal precincts, matching totalized preformed images. Such design thinking all too readily forgets its past and present, appropriating new communities as a compliment to modernity. This is the signing of a city that regulates our imagination and solidifies community as a static and transparent condition. Poetics for de-signing urban living, as an activity of deconstruction, imagines an otherwise system—one that has a failure we recognise with Blanchot at its heart. While this essay has a concern with predeterminable spacing as (place for) the locus of community, it does so only to work differently to the stasis of such community. The design here is to offer something more unaccountable, something proximal to Blanchot's *unavowable community.*[1]

Perhaps the unavowable is a space of encounter more proximal to the urban drifts of Michel de Certeau's *pedestrian utterances*[2] in the sense that the urban *fabric* is nothing less than an abundance of scripts signed by readers of the everyday, each reading the becoming-performative act of a here and now, eternally scripted. Our design thinking constitutes a performative system at work in the neighborly poetics of a community crossing Blanchot, Augé, and Heidegger. This design/writing has as its aim to bring to the fore the temporality of this performative-poetic (*unavowable*) community in a coinciding of these three thinkers.

Indeed, while there is the issue of how space is designed for activation particularly within the context of urban environments, there is still a more radical drift that I am attempting to activate as poetic. It is a drift that concerns itself with the time of community in relation to a temporality of poetics. The attempt here is to reveal how these relations find proximity across ecstatic temporalities traced in Blanchot's poetics as *politico-ethical* activity productive of a community that signs itself otherwise and a community coincident to Augé's non-place.[3] This relay or drift occurs from the perspective that community can be explored more radically than can a spatial relation of individualised selves understood as transcendental egos.

There has been a significant legacy in Western thinking that has determined our perception of spatial design, framed as the legacy of transcendental thought since Aristotle, productive of an outside conception of the world that governs human knowing. To transcend is to "go beyond" a world of inhabitation, and in this sense a transcendental tradition has set up a binary of interiority and exteriority that is still today the dominant way our spatiality is perceived. As an activity of de-signing (or deconstruction), the revealing of binaries of interior and exterior, governed by the agency of transcendental signifiers (God, Spirit, Consciousness), is a strategy for revealing belief in relation to historical change. Furthermore, de-signing foregrounds the insecurity, absence or failure of foundational terms, revealing an endless play or performance of meaning, signification or scripting that problematises fixity and closure of meaning of an image or design work. With the works of Blanchot, Heidegger and Augé, each in its own way de-signs the closure of community and opens new (non-localizable) ways of being social. Their work is not locatable according to a transcendental schema. Rather, the transcendental is always in a relation to an immanent performance of identities construed on difference, where living in the here-and-now affirms constant change and difference that everyday life exposes us to. Contrary to a set of socially and culturally prescribed and conditioned patterns that reduce identity to the same, the spatial practice of everyday communities is an inventive schema: inventing ourselves over each and every day.

We initially encounter Blanchot's *unavowable community* and Martin Heidegger's originary question of the being of *Dasein* through his notion of *ecstatic temporality*.[4] The drift then takes us to Augé's theory of urban living premised on space as exteriorized communities that have been filtered out into highly mediated zones of engagement. These zones are associated with an increase of homogenized and regulated spaces reproducible today through new technologies. In this sense communities of difference become nullified, yet our aim is to expand on this nullity or failure as the heart of difference located in what will come to be understood as *posed-solitude*. This essay maps the nullification of community as theorized by Augé. It does so through Blanchot's non-localizable community that speaks within and through his *political-ethical poetics*, only to find the inherent poetics of an otherwise phenomenon that is named here the *poetics of community*:

> Blanchot asks, in the name of the *community*, whether it is better to remain silent on what has been almost impossible to speak of without default, that is, its contemporaneous purpose. In responding, Blanchot gives and obliges his reader to not answering and at that very moment to not remaining silent. But, rather, to choose words exactingly that discover a politico-ethical poetics: "That [it] does not permit us to lose interest in the present time which, by opening unknown spaces of freedom, make us responsible for new relationships, always threatened, always hoped for, between what we call work, *oeuvre*, and what we call unworking, *désoeuvrement*" (*UC*, 56).

With Blanchot's *unavowable community* we experience impossibilities of community's existence. We face its dissolution and what cannot be pronounced in the excessive essential solitude between the "I" of a writer and the "We" that governs the relation of the work (*oeuvre*) to its own unworking (*désoeuvrement*). For Blanchot the I/We relation has the movement of nullifying subjectivity and identity in terms of what he calls the *neuter* (*le "il,"* the "he/it" or "I/we"). This is a movement of relation between the "I" of the subject (the writer) who abandons his place for the non-place of the *le il* (he/it). From this non-place the *le "il"* cannot speak, as with the speaking subject of a self, as presence and self-presence. It is a *community* that does not become another subject or identity or puts itself into dialectical opposition to the One. Blanchot suggests how this is possible:

> . . . he/it, specified as the indeterminate term in order that the self in turn might determine itself as the major determinant, the never-subjected subject, is the very *relation* of the self to the other, in this sense: infinite or discontinuous, in this sense: relation always in displacement and in displacement in regard to itself, displacement also of that which would be without place (*SNB*, ix).

This dissolution that the community of the neuter suggests elaborates Blanchot's unworking (*désoeuvrement*). *Désoeuvrement*'s association to *oeuvre* (the work of art, or literature, or spatial design) conditions a lack—or the *work* as the work's not lack—the work as unmindful of being or not-being, as neither present nor absent, inside or outside: neutral. It also means idleness, inertia suggesting a kind of re-treatment of the work that has not come to presence, that is, the relation between the work and its denial, between writing and passivity, between being and not being a writer, being and not being the subject of the verb "to write." In this sense, de-signing hides and holds from the thought of the contemporary image as in the image of programming urban sites for placing and locating identities. Blanchot's community bears a silence through the disappearance and separation of *I* and *We* in order that they come together again. In the conclusion to *The Unavowable Community*, Blanchot asks, in the name of the community, whether it is better to remain silent on what has been almost impossible to speak of without default, that is, its contemporaneous purpose. In responding, Blanchot gives, and in giving obliges his reader to not answer and at that very moment to not remain silent, but, rather, to choose words exactingly that discover a politico-ethical poetics:

> That [it] does not permit us to lose interest in the present time which, by opening unknown spaces of freedom, make us responsible for new relationships, always threatened, always hoped for, between what we call work, *oeuvre*, and what we call unworking, *désoeuvrement* (*UC*, 56).

How do we choose this path of difference where a poetics of exacting words open up unknown spaces of freedom? Something unpronounceable is disclosed here between our silence and poetics in what is described above as displacement of the self in relation to the other, whether this is in the situation of a writer to their art or a self responsible to the *polis* or politics inherent in the notion of community. This infers an acknowledgement that failure is inherent in any designing of place and that what appears in the spaces of design's failure are non-placements of shifting terrains that open onto the heart of communal difference. There is a further inference here that an other is within us already, an other we recognise as proximal to Emmanuel Levinas's notion of an *ethics* that comes before us and reduces the self to an infinite otherwise knowing of-its-self.[5]

It is crucial here to get a better understanding of the temporal order of this *be-fore* as a marker of Blanchot's displaced, non-dialectical, and unmindful relations. A key temporal register of the relation between *oeuvre* and *désoeuvrement* is revealed to Blanchot through Heidegger's thinking of time and being particularly in terms of his notion *ecstatic temporality*. However, this before of "before" will differ between Levinas and Heidegger with the latter privileging the being of *Dasein* before Levinasian ethics, which in turn comes before Being. The importance of de-signing in such temporality structured by displacement of identities, by the stranger (or other), can be traced here at a time when Blanchot's politico-ethical poetics becomes quite urgent for spatial design. Are we not living in a time when the phenomenon of displaced peoples or non-citizens (the foreigner, the alien, the stranger, the immigrant, the refugee, the criminal, the prisoner, or the outsider) is becoming the norm? We live in a time and space where no city can deny the existence of depriving citizens of their citizenship, where forced deportations, enforced migrations, refusals of the rights of asylum-seekers, where colonizations, exterminations, and exiles make up the designed conditions of our local and global spatial borders. As Giorgio Agamben cautions in his essay "Beyond Human Rights" (2003), the neologism "denizens" has been created by Tomas Hammar to account for these non-citizen residents and draws attention to the inadequacy of the concept of "citizen" in today's socio-political reality of modern states (*CWC*, 9).

However, what is most pressing here in this temporal register of *be-fore* is a priority to that of being present to itself, to a metaphysics of presence; neither absent nor present, nor dead nor alive—a wholly other, irrecoverable intrusion in our world, which is not comprehensible within our available intellectual framework whose legacies lie in metaphysics, but whose otherness we are responsible for preserving. For the displaced "denizen," the spatial designer's responsibility is recognized in holding onto all of the failures that house the wake of (Western) Imperialism, responsibility (as politico-ethical poetics) that comes before us, preserving us from mastery of, or as communal citizen identity, a preserving that is precisely the trauma of our design(ed) histories.

We approach this *be-fore* of Blanchot's politico-ethical poetic through Heidegger's *ecstatic temporality*. We must keep in mind though that in the lack of the artwork or citizen-community, solitude and excess are linked—a link constituting an excess of trauma. We will return to this significance of solitude and excess in relating Blanchot to Augé's non-place.

ECSTATIC TEMPORALITY

Ecstatic temporalilty (original time) is for Heidegger the most authentic or original notion of temporality whereby it exists primordially *be-fore* inauthentic temporality. However, this *be-fore* as my hyphen suggests is not of an order of ontic time. Inauthentic temporality can be perceived as in terms of Aristotelian time, which we understand to be a uniform sequence of now moments. What gives the being of *Dasein* is a question formulated in terms of (original or ecstatic) temporality. The radicality of this question is more than just the suggestion that as beings we are (temporal beings) within time; with-in-time (*Innerzeitigkeit*), caught within the sequence of now-time. Rather, Heidegger's suggestion is that *Dasein*'s being, understood in terms of *ecstatic temporality*, is time. Heidegger reveals the "design" of *Dasein*'s being (ecstatic temporality) in relation to its three modes of disclosure: projection, thrownness, and concern.

In projecting itself onto possibilities of being-in-the-world *Dasein* is "ahead of itself." As thrown, *Dasein* is already in-the-world; as already in-the-world it is being with entities within-the-world, in the sense that it is involved with them, dwells with them, is absorbed by them. These three temporal ecstases in their essential unity are what constitute ecstatic temporality (or original time) in terms of how *Dasein*'s being is to be. They correspond to the past, present, and future of time as commonly (or ontically) understood but cannot be identified with them. That is to say, they cannot be identified with the no-longer-now, the now, and the not-yet-now. This structure is of a different order of thinking altogether. *Ecstatic temporality* infers something being outside of time yet within it; *Dasein* is outside itself (ontically and ontologically at play). It infers another kind of movement or condition that does not conform to the logic of an ontic inside/outside binary.

Therefore to suggest *Dasein* is outside of itself, yet within it, takes a radical dismantling or *destruktion* (deconstruction) of binary thinking. What endures (the essence of being) conditions our possibility for thinking difference, or signing otherwise. This is akin to a temporal register of *be-fore* that endures and is not present to itself but is, as previously disclosed within Levinasian ethical economy, wholly other. It is an irrecoverable intrusion in our world that is not comprehensible within our available intellectual framework, yet draws attention to a necessary refrain as our acknowledgement to the trauma of (design) histories. We are responsible for preserving this otherness—formed in part by trauma—as the wholly others. It is the ghost that comes *be-fore* us, and to which we are thrown within-this-world, revealing the possibility for our concernfulness. This is a wholly-otherness proximal to Blanchot's politico-ethical poetics of community that demands a responsibility from within us, but in excess of the Self. It is not a concern we can plan for or mitigate. Rather, it is beyond the rational knowledge of an autonomous subject (Self). Modernity has failed the rights of the "denizen." In fact, the "denizen" calls on design to recognise a wholly other way: the way of a ghost who slips through time and space nomadically. Yet, to use the term "nomad" is inadequate as it is still locatable in relation to notions of stasis and nationhood. The thought of belonging conceived by the term "nomad" arises still in direct relation to a border-existence. Our attempt here is to suggest that the "denizen" deconstructs or de-signs state-nation-territory identities through being an un-representable and marginal figure that is somehow now central to our contemporary politics of urban design. Certainly, the figure of the "denizen" is one of passing or in passage, which we emphasise through Augé's transient schema of non-place.

TIME OF THE RETURN: BLANCHOT'S SOLITUDE

If we consider this marginal-central figure of the "denizen" or ghost in the Imperial trauma-machine of modernity as a more persuasive performer of our identities—construed on difference where living in the here and now affirms constant change—perhaps the *unavowable community* can be conceived as our paradigmatic practice of the everyday. Augé describes this excessive figure as the condition of Supermodernity, which he hints as an implicit poetics of self as posed-solitude. Hence we recognize with Augé something of the notion of worklessness at the heart of Blanchot's system of thought, in terms of a *politico-ethical poetics* that performs community in the margins of literature, as an ethics for a care of the present or contemporary. Blanchot's *politico-ethical poetics* opens up the unknown and performative present, conditioned by a temporality that operates between work (*oeuvre*) and unworking (*désouvrement*). His writing is the scattering of the work and of community in a movement of worklessness, which we understand as an attempt to de-sign notions of category, style, or identity construed by the proper of belonging. Augé's non-place suggests dissolution of community with respect to modernity, as a loss of individuality in relation to what he describes as the phenomenon of Supermodernity existing since the 1980s, characterized by three figures of excess: time, space, and subjectivity. Particularly important for our concerns is this figure of the lone individual who has fewer opportunities for an encounter of community as one of collective exchange. In this loss of collective-identifications, Augé suggests the opening of a critical gap in the subject between a self and self-observing-self. This he describes as a self-observing solitude or posed-solitude that marks out an implicit poetics where refrain and posturing coincide as the observation of an incompleteness that is infinitely at work, as with Blanchot's neuter or worklessness.

Blanchot's thinking on solitude, which begins his meditation on literature, shifts from the solitude of the writer in his early thought to the solitude of the work in his later writings. That is, solitude exists in the essential solitude of the work of language that is removed from the solitude of the writer alone. Or rather, in language what is to remain silent is the solitude of the writer. The paradox here is that it is in language that the solitude of the writer is expressed. Blanchot insists on this double bind where he "is not free to be alone without expressing the fact that he is alone" (*SHB*, 441).

The work of language exists prior to us and in this sense, like the ghosts in the Imperial trauma-machine of modernity, recalls our failure as autonomous subjects. The recognition of this failure is what constitutes a form of wholly other freedom. This relation, found in a freedom to be alone in solitary-self-expression encounters Augé's self-posed solitude of non-place, that he suggests is not a being-alone, but rather a being-on-our-own with the "nothing." Blanchot's solitude can only be expressed by means of that which precisely denies solitude: language. The law of writing therefore is privation whereby its other is absence and so solitude is only solitude in relation to its otherness. Blanchot suggests by this paradox: "a person who writes is committed to writing by the silence and the privation of language that have stricken him" (*SHB*, 442). The condition of possibility for literature, art, or design work is a certain silence, what Blanchot also describes as the nothing; the silence of solitude wherein the writer, artist, or designer has nothing to express. This nothing as the writer's silent solitude is the source of literature that we come to know as the *unavowable community*. More significantly for spatial design this source of our silence as the expression of nothing is the heart of difference in the non-localizable facticity of being riveted to existence without an exit, Blanchot's essential night or essential solitude. It is a ground that we re-invent—as I have suggested through such schemes as transcendental signifiers—driven by a desire for the origin of the artwork or design of self-expression. As Blanchot suggests: "having nothing to write, of having no means of writing it, and of being forced by an extreme necessity to keep writing it" (*SHB*, 442).

The radical incompletion of the artwork, literature, or design, its worklessness, is its source and thus preserves the possibility of literature/design whereby this incompletion prevents the artist standing back from her work and saying, as Blanchot notes in *The Gaze of Orpheus*: "at last it is finished, at last there is nothing" (*SHB*, 442).

Blanchot suggests a *radical otherwise knowing of community* in a performance of a *political-ethical-poetics* that gives the unconditional relation to the other. In a similar sense of an unworking community through a condition of posed-solitude, Augé believes there exists a location for new ways of being social. Here I would want to correspond Augé's potential social as posed-solitude with Blanchot's *political-ethical poetics*, through notions of excess in subjectivity as a result of the incompleteness of collective communal exchange. The self-assured identity of modernity is today faced with the traumatic gap of a past fixity of community and its now apparent incompleteness. In sensing this, Augé's gap becomes the potential translation into the opening of unknown spaces of freedom that make us responsible for new relationships not bound by collective sameness regulated by state-nation-territory. Augé's gap thereby becomes a self-performed, spatio-temporal, ethical-poetics. This coupling is a radical gesture that I hope will bring proximity to an ethics of a self that is radically opened by the other within us. It is not determinable and yet, via Augé's analysis, it would want somehow to come into closer contact with a self-performed abandonment of any *sensus communis*. Alongside Blanchot we would suggest this radical other within us breaks from any intellectual knowing, and perhaps manifests itself in the poetics of a self-performance of identity in the everyday, immanent and always under construction, fluid and transient. There the performance is possible only by its saying nothing, other than the necessity to keep on expressing itself as Blanchot's, and Augé's silent solitude between work and worklessness.

The binary inside/outside is radically sundered in Heidegger's *ecstatic temporality* and put into play through Augé's non-place as a place of excess of identities, an excess of state-nation-territory. These collective relations of identity dissolve through the transitory locales that set up sites of solitude (collective solitude) without isolation. Hence, we are not alone, but we do not disturb others. We are thinking of airports, trains, supermarkets; places of public waiting. Or there is another kind of non-place he describes as liminal or disconnected sites (grey zones) that do not properly belong to place.[6] Augé suggests that more of both of these non-places are being increasingly produced today. In terms of Supermodernity's non-places, any notion of inside and outside dissolves and becomes unworkable or incomplete.

AUGÉ'S SELF-POSED SOLITUDE

Is Augé's solitude Blanchot's? Is the self-performing or posed-solitude of the individual who is now "witness" to her loss of collective-community engaged in an ethics? How is Augé's individual with an other that is wholly other? The "individual" is not a complete entity but rather an entity haunted by the ghosts of its past (Modernity's Self). The other, today, is wholly other in the paradigmatic figure of the "denizen" who does not fit within the confines of citizen-state-nation-territory, but rather resides and brings what Agamben describes as "the originary fiction of sovereignty to crisis" (*CWC*, 8). In the posed-solitude are we engaged in a kind of poetics that can be sited as ethical and political? Does this posed-solitude implicate a form of return to modernity's autonomous self?

Augé's notion of posed-solitude is produced by a subject who is witness to her loss of community, as described by the phenomenon of the "denizen" or urban fabric woven with displaced peoples. Such peoples—the foreigner, the alien, the stranger, the immigrant, the refugee, the criminal, the prisoner, or the outsider—betray the notion of citizenship as a subject of nativity in relation to nationality. In this sense they/we are all displaced subjects today. How does this displacement become characterized through witnessing? This notion of witness to a loss of community is not an act of self-consciousness as in the subject witness to proof, but rather it is a subject of testimony. Testimony as posed-solitude becomes a performative utterance through the prosthetics of subjectivity caught up in the archiving, recording, rehearsing, retelling, the sending and receiving of ourselves: our futures. It is a time-out-of-joint where the present experience defers in the rehearsal of ourselves alone, in solitude but not isolated; a type of collectivity without isolation. In non-places we are not alone, but we do not disturb people.

As Augé suggests, these performative utterances that occur via the prosthetics of the subject caught up in archiving and rehearsal, are activated through the way we are in spaces with our technological devices: cell-phones, MP3 players, Ipods, I-phones, laptops, and digital cameras. Augé, in his text, examines an array of spaces productive of a transitory and waiting nature, such as the supermarket, the ATM, and, most significantly, the airport. In our being in non-places Augé critically observes a gap that opens up between a self and a self-observing-self. The experience of travel is the example he nominates for a deeper engagement with this observation. Through the provision of excessive memory in digital cameras, the documentation of our travel extends further into a self-securing activity that can produce images of ourselves and spaces en-route as much as the destinations to which we travel. Augé suggests that this overly mediated activity of experiencing place produces a situation where we are today more-than-ever engaged in a heightened process of rehearsal and retelling for the future. That is, we have projected ourselves into a future of our telling of the journey encountered, as in a "this will be great for the future" archive.

This future retelling is not new, as it has existed for some time before the advent of photography as seen, for example, in the art of portraiture painting. We could even suggest it is constitutive of every teleology, as such, every narratival structure and every story-telling event. However, Augé's point is that it is a more heightened phenomenon as a documentation of the self, *by* the self in and as posed-solitude. This time-out-of-joint—a Derridean expression inherited from Heidegger's *ecstatic temporality* by way of Shakespeare's ghost of Hamlet—is for Augé a temporality of the double gap of an uncertainty for the future that is both excessively predetermined or predestined, and that never arrives. Or, to put it another way, there is an arrival at the destination before even having taken the journey; an arrival that defers and differs from that which was imagined once. We recognise in this the temporality of Derrida's *différance*, Heidegger's *ecstatic temporality* and Blanchot's *récit*.

RÉCIT: FRAGMENTS AND STEPS (*PAS*, STEP-NOT): *THE MADNESS OF THE DAY*

I want to focus on two particular moments in Blanchot's writings: the first is formally a *récit* that hints at an end to all such formality in *The Madness of the Day*; and the second is a form of criticism in *The Gaze of Orpheus*.[7] What is this end to formality? The movement from *roman* (novel) to *récit* (recounting) appears initially in Blanchot's two versions of *Thomas the Obscure* to the refinement and eventual disappearance of the *récit*; what is called in *The Madness of the Day*, the *pas de récit* (the one step more/ no more of the tale), when Blanchot stops writing "fiction" altogether (or so it seems). Both Blanchot's fiction and criticism reach a point where they undergo fragmentation and pass into one another, something that can be seen particularly acutely in *The Writing of The Disaster*.[8] I suggest that we would want to read Blanchot's work as a movement toward a kind of transcendental immanence from the distinction between fiction and criticism and form and content implicit in both genres. But Blanchot's transcendental immanence is a gesture toward a spatial design tactic. That is, it is a gesture to rethink and de-sign the concept of boundary. The urgency here is to design through an abandonment of ownership and sovereignty, where even "public" and "private" become unconvincing spatial concepts. De-sign is the tactical embrace of the Blanchotian neutral as fragment. And further, we might see this Blanchotian example as the production of literature as its own theory, and through its genre of expression it is the fragment. This fragmentary position transcends comprehension in its refusal and produces an alterity irreducible to presentation or cognition. This is a wholly other that can be named variously: *absence*, *the essential night*, *community*, *silent-solitude*, *radical passivity*, and *worklessness*. In an improper fashion, then, I suggest that Blanchot's *récit* is a site for excavating Augé's temporality of a double gap that marks out an uncertainty for the future that never arrives.

Blanchot's *récit* activates a space of literature where time breaks and another order of time takes—not in the sense of a taking for mastery, for possession or centrality or for comprehension. Rather, this situates a temporality of taking that expresses Blanchot's writerly concern in the "form" of the *récit* as an exhaustion around the mere condition of possibility of narrative and in this same moment its impossibility. For spatial design's orthodoxy, the fiction of community identity operates as a foundational scene for the planner. We would suggest the urban fabric is composed of performative scripts that de-sign the stasis of design's image. Take the "short cut." As it cuts its way through an urban fabric it interrupts the programmed syntax of spacings (as place-making). The cut-up text as word-salad hints at the transgressional nature of spatial intervention in the conventions of a novel. Blanchot takes this significantly further in relation to the *roman* and *récit* just as the denizen-de-signer pulverises the familiarity of a city topology. In *The Madness of the Day* the narrator recounts, that is, he gives us the possibility of narrative, of a story of his existence that does not add up to those who see him otherwise. The other as Law of Reason cannot understand why this man has ended up where he has when he had so much promise. They see a promise, a potential, because he is a story-teller and yet, these representatives of the Law cannot understand why a man who has such ability in telling cannot recount an orderly narrative of his own existence. Likewise, the protagonist/narrator confesses this is beyond his capacity: "I had to acknowledge that I was not capable of forming a story out of these events. I had lost the sense of the story" (*SHB*, 199). Blanchot's protagonist/narrator activates the impossible-possibility of narrative by frustrating the story in starting it over-again at the end.

This is to suggest narrative is a recounting of experience that knows the experience only through a retelling. Here, *Augé's* self-posed-solitude is not far from such an experience. All narratives are a rehearsal, an archiving, a retelling or repetition as iteration. Narrative impossibility (and possibility) is in its "recounting facts that he [the writer] remembers" (*SHB*, 199). The remembrance is another story in terms of why he might be retelling it at all. I have suggested earlier that this is linked to an insatiable desire for the source of the artwork and design as worklessness.

The radical incompletion of the artwork, literature or design, its worklessness, is its source and thus preserves the possibility of literature/design whereby this incompletion prevents the artist standing back from their work and saying, "at last it is finished, at last there is nothing" (*SHB*, 357). The notion of design as completion is a ruse of identity as an autonomous narrative. Rather, the work of de-sign reveals the failure of a complete community at the heart or source of our desire for invention.

Blanchot's *récit* enters into the space of literature where no more stories happen as they depart from their sense. This space of literature, Blanchot's ethico-political-poetics, or the space of de-sign—between the possibility and impossibility of narrative or *récit*—activates this relation between work and unworking. The temporal logic (or madness) of Blanchot's *pas* (step-not) is of the story "of what never happens," or the impossible narration that is the whole story of the non-story of Blanchot's *récit*. This temporal *pas* (step-not) is conditioned by the impossibility for the narrative to continue too rapidly in order to give into the demands of a metaphysics of presence; into the clear and direct light of day.[9]

TIME OF EXCESS: SUPERMODERNITY AND *THE GAZE OF ORPHEUS*

It is precisely the temporality of starting over, or perhaps, of never beginning in the wake of deferral that marks Augé's new way of being social within (t)his phenomenon of Supermodernity. This is a phenomenon that locates excess within posed-solitude of a being-with-ourselves otherwise. As mentioned, Augé's Supermodernity is marked by excess in the three figures of time, space and subjectivity. Time is excessive in the sense that Augé notes we are now experiencing more events with more frequency, as more people exist in increasingly mediated societies. We are drowning in events, which fill our time in an intensive way. We live longer which means there is an increase in the generations alive at the same time whereby history is at our heels in the sense that we experience first hand those who have been through historic events (wars, etc.), increasingly so. Augé suggests we have a difficulty with time in the sense that with an excess of meaning time contracts; there is a density of the present. In the same way time is experienced, space too is multiplied wherein we experience spaces through a variety of ways not only mediated but also scaled and multiplied through the different perspectives we have of a space. The cultural significance Augé suggests lies in the distinction between cultures that would have created myths, religions, or cosmologies to make sense of their world and those new-technologies, which give us super-human access to (techno-scientific) ways of knowing the world. He argues we have a lack of mythology in the world due to the excessive mediation of the world, which now explains too much. Our facility to view the world via multiple spatio-temporal registers drains a kind of questioning that is perhaps more primordial, more originary.

I am particularly concerned with Augé's subjectivity-in-excess that is productive of a double gap between the self and self-observing-solitude and posed-solitude as a performance. This structure of the double gap locates my proposition for a possible poetics of solitude shared within this condition known as Supermodernity. This loss of a *sensus communis* has produced an excess of belonging-not-belonging in the space of rehearsal, retelling, archiving, being-alone-with-oneself, what we would want to call a poetics of the self-in-ruin. In Blanchotian terms this would be the site of the *unavowable community* of "the never-subjected subject as the very relation of the self to the other" (*SNB*, ix). In this sense infinite or discontinuous, a relation always in displacement and in displacement in regard to itself, displacement also of that which would be without place (non-place).

The suggestion here is that an excess of our subjectivity maps out a relation between the self and other-to-(or within)-our-self revealed conditioned by the techno-prosthetic epoch of supermodernity. It is self-posed-in solitude that marks an excessive desire or rather, as Blanchot would suggest, desire is the condition of excess always in excess of law. It is through Blanchot's critical engagement in *The Gaze of Orpheus* that we locate this self-posed solitude, that marks Augé's double-gap between a self-observing-solitude and a self-posed (or perhaps possessed) solitude that defines an excess within a lack, as Blanchot's source of the artwork (our design); the origin of community's productive force through anxiety. Here desire and angst are the most productive forces for work and worklessness.

Writing engages a movement toward the nothingness opened by the experience of dread or anxiety. Literature is an attempt at saying nothing; dread is nothing expressible and yet the only thing that causes me to desire expression. Writing is useless and yet nothing is more serious. We must return to the theme of solitude to make clear that Blanchot's freedom or autonomy of the writer in the privation of language "is not free to be alone without expressing the fact that he is alone" (*SHB*, 437). This is an autonomy that can never achieve complete self-identity through the alterity of the artwork. Blanchot makes the distinction between essential solitude and the solitude in the world (*SL*, 251–53). Essential solitude is not the worldly, artistic solitude as in the mythic artists alone. Such solitude for Blanchot is existential solipsism, which is self-relation or self-communion. Rather, the essential solitude is that of the Work, a solitude upon which the writer is dependent but to which she necessarily has a self-deceptive relationship, mistaking the Work for the Book she writes. This is, in short, a totalized view of the artwork. At stake in *The Gaze of Orpheus* is the conflict between the law of the artwork and the lawlessness of desire that exceeds this totalization of Work. What is clearly mistaken in Orpheus's *gaze* is the relation between the work and the capital W-Work as a totalized form: it is the moment when Orpheus turns around to look at Eurydice in the night, as the night he transgresses the law (of the underworld) through the movement of desire. For Blanchot, desire is always in excess of the law. Orpheus's desire is not to see Eurydice in the daylight, in the beauty of a completed aesthetic form that has submitted to the passage by way of the law of concealment, but rather to see her in the night, as a figure of the night prior to daylight, "her body closed, her face sealed" (*SHB*, 438). Orpheus does not want to make the invisible visible, but rather (and impossibly) to see the invisible as invisible. Orpheus's "mistake," as it were, lies in the nature of his desire, which desires to see Eurydice when he is only destined to *sing* about her. He loses her through his desire and is forced to forgo both his art—his song—and his dream of a happy life. The paradox of Orpheus's situation is that if he did not turn his gaze on Eurydice he would be betraying his desire and thus would cease to be an artist. Thus, the desire, which destroys his art, is also its source.

This ambiguous locale of source and destruction or the work's failure is located within the Blanchotian notion of work and worklessness. Orpheus's gaze traces out this ambiguity that we have attempted to locate or at least open up in the space of Augé's double-gap between a self-observing-solitude and a self-posed or perhaps possessed solitude. That which possesses the self in this excessive zone of retelling, rehearsal, acting-out of self is the gaze of the other from which our poetic-desire is sourced and exceeds us. And further, what drives us to keep producing "new" work, whether this be the constant retelling of ourselves within our self-observing solitude or posed-solitude, is what Blanchot terms community or the "other night," that is a dying stronger than death in our self riveted to existence. This energy and desire of worklessness, or "the other night," excessively works against the law of metaphysical truth fuelled by an energy of the lawlessness of writing's desire as originary difference (Derrida's *arche-writing*). Blanchot's *le mourir* (or other night) is the stronger night that gives the origin of the writer's experience as the impossible experience to control one's death. Further, this origin is something stronger than death, namely the simple facticity of being riveted to existence without an exit. The desire that governs Blanchot's work, and what I have tried to express in Augé's self-posed-solitude, has its source elsewhere than from the dialectical movement of self-consciousness. This experience of *le mourir*, the essential night, where one cannot find a position, is the experience of the *other* as source of *nothing*.

NOTES

1. Maurice Blanchot, *The Unavowable Community*, trans. Pierre Joris (Barrytown: Station Hill Press, 1988). Hereafter cited as *UC*. If the question of "community" poses a radical drift here it is not a community of the Self-Same constituted by an identity that would provide a locale. Rather, it would be an impossible community, an "inoperative community," a community constituted on the spacing of difference. See, also for example, Jean Luc Nancy, *The Inoperative Community*, ed. Peter Connor, trans. Peter Connor et al. (Minneapolis: University of Minnesota Press, 1991).

2. Michel de Certeau, *The Practice of Everyday Life*, trans. Steven Rendall (Berkeley: University of California Press, 1984). Hereafter cited as *PEL*.

3. Marc Augé, *Non-Places: Introduction to an Anthropology of Supermodernity* (London: Verso, 1995).

4. Martin Heidegger, *Being and Time*, trans. Joan Stambaugh (Albany: SUNY, 1996).

5. Levinas Emmanuel, *Totality and Infinity: An Essay on Exteriority*, trans. Alphonso Lingis (Pittsburgh: Duquesne University Press, 2001/1969).

6. "Grey zones" are spaces or places of alterity. They could be Marc Augé's "non- places," Michel Foucault's "heterotopias," or Edward Soja's "thirdspace" (just to name a few). They exist as real spaces and places we know and are also new spaces created by the use of technology. See Marc Augé, *Non-places: Introduction to an Anthropology of Supermodernity* (London: Verso, 1995); Michel Foucault, "Heterotopias," in *Rethinking Architecture: A Reader in Cultural Theory*, ed. Neil Leach (London: Routledge, 1997); and Edward Soja, *Thirdspace: Journeys to Los Angeles and Other Real-and-Imagined Places* (Cambridge, MA: Blackwell, 1996).

7. Blanchot, Maurice. "The Madness of the Day," in *The Station Hill Blanchot Reader, Fiction & Literary Essays*, ed. George Quasha, trans. Lydia Davis, Paul Auster and Robert Lamberton (Barrytown: Station Hill, 1999). Hereafter cited as *SHB*, 189–200.

8. Maurice Blanchot, *The Writing of the Disaster*, trans. Ann Smock (Lincoln/London: University of Nebraska Press, 1995/1980).

9. Maurice Blanchot's *The Madness of the Day* is a short *récit* that has been discussed by Jacques Derrida extensively in "The Law of Genre," and also in "Living On/Borderlines," in relation to Blanchot's *Death Sentence*. What for Derrida is hidden, suggested already, is the temporal gift of the not-yet, the Law's hiddenness as abyssal difference in the story "of what never happens." The law in Blanchot's story appears as a *feminine* "silhouette" that is neither a man nor a woman and is a companion to the quasi-narrator who is before the law. What is impossible to narrate is the story of the law, an impossible story recounted and demanded by the law's representatives (policemen, judges, doctors). The story recounted, that is "put forward," as appearance, to the representatives is on the impossibility of recounting as correctness, as presence, and hence its impossibility. Derrida suggests this union of an *impossible story* or story as the *impossibility of possibility* is where literature begins. It is made impossible before the representatives of the Law ("language is the elementary medium of the Law"). This is the union bringing together an "I/We" of the "remarkable truth" of truth as more adventurous and risky. At that point it would be a truth without end, abyssal, as random drift. Yet, more significantly, it is the "I/We" not of its representatives, but of the law *herself* who, throughout a *récit*, forms a couple with me, with the "I" of the narrative voice. Further, as we know not *what* or *who* the law *is*, as in the neutrality of its non-gendering, the law opens up the impossible "atopology" that annuls oppositions. For further reading see, Jacques Derrida, "Living On/ Borderlines," in *Deconstruction and Criticism*, trans. James Hulbert (New York: Seabury Press, 1979) 75–176; and Jacques Derrida, "The Law of Genre," in *Acts of Literature*, ed. Derek Attridge (New York/ London: Routledge, 1992), 221–52.

II

Geo-Placed Knowledges and Design

Geo-Placed Knowledge

Harriet Edquist

The underlying thread that ties together the chapters in part II is to do with how we might map different geographies. A key issue revolves around the application of map-based information systems and thinking to non-conventional areas. An additional issue focuses on how we represent space and place through the mediation of designed practices. The collection as a whole examines the intersections between old and new technologies and between design and other modes of practice; it brings to the fore issues to do with the constructed interfaces between the viewer and the world in our ways of seeing. Part II is concerned with the idea of mapping as a way of negotiating complex geo-spatial information, of how we find our way in the world and of our understanding of place. The ways in which each author in part II addresses these issues vary; in the first chapter, apprehension of space and place is investigated through the medium of a design project; in the subsequent chapters the authors critique the affects of the representational practices of cartography, ethnographic photography and film, and the literary text.

For Laurene Vaughan and Yoko Akama our apprehension of space and our construction of place became the subject of a design project, which they undertook with a group of undergraduate students from RMIT University. This design intervention, called the "24 Hours Noticing Project," took place in Hobart, Tasmania. Here two different ways of understanding place came together: the Western *practice of noticing* and the Japanese concept of *Kū*. The objective of this conjunction in the studio was to explore the relationships between our perceptions of space and our experience of place through the experience of journeying, and to determine whether, or not, this experience can inform design practice. While "noticing" and *Kū* are quite distinct ways of apprehending the world, their conjunction in the design project brought to the fore the ways in which designers negotiate location in order to create something new. From this project, further questions emerged to do with the ontology of design; they were pursued using both European and Japanese theoretical and conceptual frameworks.

William Cartwright argues that the traditional, scientific model of representing geographical space is inadequate to represent the complexity of the observed world and that this model was developed under the specific professional regime of the map-makers. Paper-based maps, while at times objects of great beauty in themselves, can restrict or mislead the viewer's apprehension of a place, but the addition of other media, such as three-dimensional representations, no matter how crude, can increase the apprehension of "reality." Cartwright infers that the great age of cartography was one where the mapmakers worked in concert with the new printing technologies, which produced their maps, and that today, with the proliferation of new digital media where content and means of production are again in concert, we have the opportunity to enter into a new age of elegant and rich map-making. To achieve this however requires collaboration between disciplines across the science and technology spectrum and the design-arts spectrum. By exploring the development of the map, Cartwright raises questions about the affective nature of design and the relationship between seeing and knowing.

In her examination of the way Rolf de Heer and Peter Dijgirr restaged Donald Thomson's historical ethnographic photographs in their feature film, *Ten Canoes* (2006) Linda Daley raises issues to do with new ways of encountering the colonial subject and the role of mimesis in constructing and reconstructing identity and place. She explores the relationship between the Thomson archive of photos and the reconstruction of the cultural practices of the Yolgnu of Arnhem Land. She argues that while post-colonial critiques of ethnographic photography usually find the practice exploitative, in the case of *Ten Canoes* this was not the case. Indeed the materiality of the photograph itself provided an interface or medium through which, by looking, touching, stroking, Yolgnu were able to reconnect with their ancestors, mimicking ancient practices. This gives a weight to the material artefact that we forget at times, accustomed as we are to the transitory nature of the visual image, which can be displaced/replaced through so many different media. Through the experience of filming *Ten Canoes* however, the Yolngu "recoded" the Thomson images through their bodies as actors, which, as Daley points out, is possibly "the most intensive form of affective encounter possible with a photographic image." In this instance, as in the "24 Hours Noticing Project," the cultural assumptions of Western "ways of seeing" are brought into question.

In the final chapter, Harriet Edquist examines the ways in which characters encounter space and construct place in a nineteenth-century colonial text set in Melbourne. In her study of Ada Cambridge's *A Woman's Friendship* (1889), Edquist uses the emerging discipline of literary geography to provide a spatial reading of the narrative and to provide new ways of paying attention to hitherto unheeded aspects of the novel. Normative criticisms of *A Woman's Friendship* tend to focus on the plot of the novel and the characters of its two female protagonists. Edquist, on the other hand, offers a counter argument by focusing on the underlying spatial narrative, particularly the domestic interiors where much of the action takes place, providing an alternative reading to one focused solely on plot development and character.

Chapter Six

24 Hours Noticing

Designing our Encounters with Place

Laurene Vaughan and Yoko Akama

Many different variables inform our way of knowing, experiencing, or encountering space. Cultural practices, personal or collective history, and social norms inform our encounters with space. These encounters are framed by their contexts and intentions, for example, the intentions for being there, and whether the encounter is solitary or communal.

KNOWING AND ENCOUNTERING PLACE

To know a place is a unique human experience. Our expectations of knowing, or even just our ability to know, will inform the depth of our knowing. Encounters with space and place occur during the trajectories of everyday actions, whether we are at home or away, and such encounters can be undertaken and experienced by anyone. One person could claim a knowing of a place by seeing it on the television, or passing through it regularly on a drive to work. Yet another person may consider knowing the same place to be a surface level of engagement even after having lived in this place for twenty years, and acquiring knowledge rich with the nuances of geography, history and practices of habitation.[1] Beyond the everydayness, many of us seek out new encounters with space as a means to extend our understanding and experience of the world. Undertaking the grand tour was once the final component of a rounded education. Travel for many people continues to be an important source of creativity and inspiration. Founded in this perception of the evolution of knowledge and knowing, *24 Hours Noticing* drew on the experience of journeying as a method for discovering the experience of place. There were several purposes behind the project: firstly, to design encounters with space to heighten awareness of space and locale, and secondly, to design a means for communicating the outcomes of this to others through a digital narrative of place. The project exploration provided a fruitful ground to explore and inform practices in design.

24 Hours Noticing was a design project that explored the potential synergy be-tween two culturally different approaches to engaging with the experience of space and place. These were the *practice of noticing* and the engagement with *Kũ*. One is Western, the other is Eastern in origin; they are similar and yet different, and their combination provided us with a way to conceptualize and communicate our experience of the inhabited world. John Mason proposes the practice of noticing as a way through which we are able to understand and create a relationship to place.[2] He argues that the practice of noticing can facilitate awareness, reflection, learning and transformation.

Kũ is the second approach in this project. The literal interpretation of *Kũ* in Japanese is "space." According to many Zen Buddhist texts, *Kũ*, is often associated with *Mu*, meaning "emptiness," "non-being," or "nothingness." For example, Kasulis, who has published several books on Zen Buddhism, discusses *Mu* as "without thinking," a pre-reflective mode of consciousness as the very ground of immediate experience. He explains that for the Zen person who operates in such fashion, "experience is grounded in its most direct contact with concrete reality."[3] Though these concepts of *Mu* are interesting areas for discussion, this chapter does not use *Kũ* associated with *Mu* and Zen personhood, rather *Kũ* is interpreted as discussing the agency of space. In this project we engaged with the interpretation of *Kũ* as a space of "potentiality" rather than its often-associated meaning of "emptiness" or "nothingness." Emptiness or nothingness could be understood as encompassing the potentiality of space, where the potential is generated through the absence, the unexpected, or the unfamiliar. Generating potential through an exploration of the unknown is a familiar activity in design. The designer, who explores the act of knowing the unknown through an object or image, must embrace the limitations and challenges encountered during the design process. To practise design in this way is to step outside of ideas of certainty and to embark on an exploratory path of discovery. This was essential to the *24 Hours Noticing* project. The selection of different localities and landscapes of the project held potential and inspiration—*Kũ*—for each participant. As they traversed their multiple yet individual paths of noticing, a particular practice evolved—one that resulted in a collective of individual views that were visually expressed as disparate things.

Engaging with *Kũ* through the practice of noticing requires one to surrender to the unknown and to uncertainty. To "notice" is not to seek out or to scrutinise, for as soon as one tries to look for or strategize a plan for focusing, genuine engagement with *Kũ* is dissipated. Instead, noticing draws our attention through our peripheral vision. Like a glimpse, it enables us not to look for, but rather to chance upon "something or someone serendipitously" (*ROP*). This approach and engagement with *Kū̃* and its application to design practice may be considered contradictory to common definitions of design. "To design" is often interpreted as "to plan" or "to provide a description."[4] Designing is often perceived as a way to fulfill a plan and to provide a description to the client and users of what is to be expected in the design outcome. However, in this project's discussion, we explore design as a way of engaging with space, a means for enunciating the unknown, or a way to create meaning from the abstract; in the same way that we can conceive of noticing as a temporal practice of discovery and place-making. Participation in the project required designers to step out of their normative practices and inculcated definitions of design. This release of control and stepping into the unfamiliar and uncomfortable territories is highly confronting to the designer. However, the rewards for this risk are in its potential to stimulate the discovery of the unknown that a more predictable process may not offer. As one walks through the landscape (urban or rural), the interweaving paths of our trajectory give shape to spaces, and thus create an awareness of "self" as the body moves and transitions from place-to-place.[5] Thus through noticing and *Kū̃* our experience transitions from the ambiguous openness of space into the connectedness of place (*TPE*).

DESIGN PROJECT: 24 HOURS NOTICING

24 Hours Noticing was a project undertaken by eleven communication design researchers in May 2007. The catalyst for the project was the group's visit to Hobart, the port capital of Tasmania, Australia. The brief for the project asked the participants to explore the practice of noticing as a method for experiencing and creating a sense of place. A 24-hour timeframe was set as a limitation within the project brief and participants were to document their transitions through this location. Prior to embarking on their individual adventures, the design researchers were given the following piece of text intended to provoke their thinking about the practice in which they were asked to engage:

> Noticing is an act of attention, and as such is not something you can decide to do all of a sudden. It has to happen to you, through the exercise of some internal or external impulse or trigger. The more you notice, the more you can accumulate to support noticing in the future. Marking is also an act of attention. It involves attaching connections so that what is marked can come to mind later without the need for outside triggers (*ROP*, 61).

Mason's text guided the participants in the practice of noticing as they documented the surrounding spaces, places, signs, and landscapes. The group often visited particular locations together, which fostered a community experience that resulted in a level of shared and common observations, and documentations. This activity combined a practice of noticing as an individual act informed by the noticing of each other's noticings.

Over the twenty-four hours, a diverse visual collection of over three hundred images was amassed between the design researchers. Although the documentation was the outcome of individual observations, the images had an echo or trace of the presence of the others and their shared experiences. These images were later curated as an on-line exhibition by two of the project participants. The challenge for the curators was to create a visual narrative that represented the diversity of observations while also enabling the viewers to notice for themselves as they experienced the outcome; what we might call a "second order" of noticing. The curators wanted to retain a balance between the individual and the collective, as this was an experience and a perspective that was central to the project. The screen display was driven by a database, where images repeated and were cropped randomly through a gridded frame and overlays. This curatorial approach generated a serendipitous encounter that allowed the process of noticing to continue, even in its second order of meaning. Noticing became a method of creating and also a way of engaging the viewers of the exhibition.

ENCOUNTERING THROUGH DESIGN

The experience and reflection of the *24 Hours Noticing* project was catalytic in raising questions about how design is defined, and how designers can engage with their world through design. The dominant paradigm of design that has emerged in Western theory and practice is one that privileges design as an active and conscious mode of engagement. Much of this is reflected in the numerous published texts that focus on what design *does* in and to the world as opposed to emphasising what it means for design and the designer to *be* in the world, which significantly shifts our understanding. Design research is often characterised by learning more about what design is; its materiality, its impact, its methods and processes for involving and engaging people in its production and outcome. The design research community acknowledges that this area of scholarship makes a valuable contribution. However, some have also questioned the omission of an ontological way of understanding design in this discourse. Tony Fry[6] and Anne-Marie Willis[7] both seek to perceive design as a subject-decentered practice rather than one obsessed with objects. According to Willis, ontological de-signing differs from the predominant paradigm of design as it seeks to know "how we 'are' and how we come to know who/what we are" in this world (*OD*, 1). We design this world, which in turn acts back on us and designs us—a process that results in us being designed by our designing.

The dominant design paradigm privileges an active, logical mode of discovery and creation, grounded in Western language and shaped through contemporary education. This language (a singular term for a range of Germanic and Latin based outcomes) institutionalizes a mode of engagement that emphasizes analytical intellectual facility. It is a mode of consciousness that favors the active, physical mode of experience that results in selective perception.[8] Henri Bortoft explains that human beings have two major modes of organization: the action mode and the receptive mode. The action mode refers to a consciousness that discriminates, analyzes, and divides the world up into objects. In relation to this dominant understanding of design, the action mode is aligned with descriptions and discussions of what design *does* to the world. In contrast, the receptive mode is best described as openness, for example being open to events as they happen. It is an alternative mode of organization that utilizes holistic, non-verbal, non-linear, and intuitive modes of communication. It emphasizes sensory and perceptual consciousness, and is based on taking in and working with what is, rather than manipulating an environment or situation to some predetermined outcome. Positioning design as a receptive mode of engagement enables design to be perceived as a method that draws on and deepens the designer's ability to understand what it means to *be* in the world within the context of things as they happen. This mode of engagement requires a way of designing that is open and receptive to the world, rather than one of making things fit a pre-planned outcome.

Bortoft argues that in order to reverse the way in which we engage with the world from one that focuses on an analytical, sequential, and logical mode of consciousness, one must turn one's awareness from the singular *object* and encounter the *whole*. Using Goethe's science as the basis of his argument, Bortoft states that recognizing and distinguishing one thing from the other immediately separates oneself from the thing; it requires that we stand outside of it. This mode of consciousness implicitly limits the possibility for us to experience an authentic wholeness. "This turning around, from grasping to being receptive, from awareness of an object to letting an absence be active, is a reversal which is the practical consequence of choosing the path which assents to the whole as no-thing and not mere nothing" (*WN*, 17).

Bortoft's concept of an "active-absence" or the whole as "no-thing" is complex and paradoxical. In order to build on Bortoft's notion of experiencing an authentic wholeness and apply it to the practice of design, we turned to using Japanese language and concepts to argue the main ideas of this discussion. The Japanese language is conducive to articulating and capturing symbolic, abstract notions of perception and experience, especially with regards to ideas of "absence," "nothingness," or "emptiness." These words are of particular significance to our interest in exploring ways and implications of people's encounters with space, encompassed in the *24 Hours Noticing* project. Western semantics, based on logic and consciousness, can often emphasize and subsequently heighten awareness and perception that something is missing. However, the Japanese language, which borrowed many concepts from the Chinese language, has evolved over time through the influence of Zen Buddhism. This evolution has facilitated a language and mentality that is able to conceptualize, as well as articulate, notions of "space" that is not understood as an "absence" of things, nor is it "nothingness" or "emptiness" where the focus is on something that is *missing*. *Kũ*, acknowledges the existence and perception of an active force that occupies its conceptual, physical or time-durational dimensions.

It should be noted that meaning in this context is not what may be perceived of as deep meaning, which is the kind of meaning that emerges from a deep knowing or mapping of a place[9] in order to know about it. Rather, our use of meaning is a personal level of connection between the self and the object or context of attention. This interpretation of space allows for a dimension with agency. It is a productive space, not a void to be filled. Its notion is similar to the pauses and silence in music—an active presence in the totality of the musical experience.

The *24 Hours Noticing* project provided an opportunity for the design participants to develop an awareness of the concept of *Kũ* through an engagement with the practice of noticing. However, the resultant images from the *24 Hours Noticing* project do not capture *Kũ*, as *Kũ* is not a space that can be contained or rendered through an image or object; in fact such actions and outcomes destroy it. *Kũ* ceases to *be* when captured, documented or described. This transformation of becoming an object "thing," like a photograph depicting a subject, only reveals a defined physical entity through a chemical processes. Although it is possible to argue that *Kũ* could exist within the photograph as a space of possibility, manifesting through the gaze and interpretation of the new viewer of the image, this space of reading/viewing is not the same thing as the image itself. This act of perception is with the being that perceives; it is not in the thing itself. The embodied action of perception, this encounter, is a different space that could be interpreted and engaged with in many ways, according to what the viewer brings to the occasion.

EMBODIED PERCEPTION

Integral to both the experience of *Kũ* and noticing are practices of embodied perception. These are means of encountering space that engage the whole being. It is intellectual, emotional, and sensorial. Each of these modes of knowing and being in the world, whether engaged consciously or not, support us in deeply knowing the world we are in. The body is one way of articulating this integration, but it is not just a physical entity of flesh and bone; rather, it is a multi-modal vehicle for transforming the world. The objectives of both of these practices (*Kũ* and noticing) are to provide us with a more meaningful and connected understanding of the world. They are agile methodologies that, in their ease, enable us to see what might otherwise be lost or unnoticed. We are not bound to a need to structure, rather we are free to *be* with the things we encounter, and from there, form a relationship. As argued by Malpas, this is essential in the creation of a sense of place and what it means to be human in that context.[10] "There is no possibility of understanding human existence—and especially thought and experience—other than through an understanding of place and identity" (*PE*, 75).

One of the key challenges of embarking on a *Kũ* informed exploration of place is that the agency of *Kũ* requires that we engage with each experience anew. *Kũ* does not rely on any accumulative understanding of life and the world.[11] Like *Kũ* itself, we must become a space ready to receive; we must see each thing afresh; it is an approach congruent with the notion of the peripheral view or glimpse of the practice of noticing. To look is to know what you are seeking before you start. To notice is to chance upon without expectation. Such an approach to exploring a location, familiar or unknown, opens up possibilities to be somewhere that you never expected to be.

Engaging with *Kũ* through the practice of noticing can enable a transition from abstraction to meaning. This practice of noticing brings into our consciousness the elements of our environment (tangible or ephemeral) that may go unseen. There is something about these elements that allows them to be noticed in some way that would otherwise be passed by in a sea of grayness. It glimmers and seduces, makes itself known. As such, our perception of these unfamiliar, foreign, or unseen elements, those which are outside or beyond the self, transitions from being outside of the self to becoming part of the scope of possibility. This transition is purely one of perception; nothing has changed for the thing that is being noticed. There has been no action, no statement; this shift in meaning is purely in how it is perceived.

The *24 Hours Noticing* project sought to explore different modalities of conceiving people's encounters with space through a focus on one particular site. Participants were asked to do this without influence of previous knowledge or expectations and to utilize Mason's definition of the practice of noticing as a methodology of happenstance. The process involved an implicit, embodied, and subconscious balancing act between what could be seen with a naked eye, peripheral vision, and viewfinder. These physical shifts could only occur through a shift in their cognitive and perceptual methods for being in space. The project brief was the means to highlight the presence of the two key elements that would otherwise be lost in their everyday familiarity—the body and the landscape/place that it is in. For as Casey states, these two elements are so ingrained in our experience that they go unnoticed for the most part (*TPE*).

The ongoing artefact of this design exploration is a curated digital exhibition (http://noticing.collabo.net); a digital physical record of a narrative layered with many meanings. Even though it is an outcome of the *24 Hours Noticing* project, the curatorial design of its form is a second order articulation of the original brief of the project. It is more than a record or an exhibition of the participants' images. Constructed as a random database-driven narrative, the outcome provides the opportunity for people to notice, and for the participants to notice once again. This digital narrative is a space that holds the potential for noticing, and to notice what was noticed, and to notice the relationships between the images of individual noticings. Similar to a hall of mirrors where the relationships between image, object, and reflection can manifest a web of knowing, it is the layers of relationships that enable the view and subsequent meaning making.

If *Kũ* was present in the initial acts of noticing, can it still be present in this curated, designed state? Can a viewer of such a narrative, be looking at a space of visual communication with the elements of agency and potential that the physical world once had? Can its visual projection and its interactive system be a space ripe with possibility? The answer may be, possibly, yes; but not in form. It can potentially become *Kũ* through the viewers' perspective and expectation. *Kũ* is a way of being in relation to space. It is a way of viewing and experiencing. In this way, the design intervention is not in the *thing* that is made; rather the design intervention is a *way of being* realized through an encounter with a particular space. Such a perspective calls for a radical shift in the understanding of design; the focus is not on the design, the thing that is made. Instead the focus is on that which is constantly in a state of making, through the eyes and experience of the perceiver.[12]

NOTES

1. Eamonn Wall, "Walking: Tim Robinson's Stones of Aran," *New Hibernia Review* 12, no. 3 (Fómhar/Autumn 2008), 66–79.
2. John Mason, *Researching Your Own Practice: The Discipline of Noticing* (London: Routledge Falmer, 2002). Hereafter cited as *ROP*.
3. T. P. Kasulis, *Zen Action/Zen Person* (Honolulu: University Press of Hawaii, 1981), 100. Hereafter cited as *ZAZ*.
4. Nigel Cross, *Designerly Ways of Knowing* (London: Springer, 2006).
5. Edward Casey, "Body, Self, and Landscape: A Geo-philosophical Inquiry into the Place-World," in *Textures of Place: Exploring Humanist Geographies*, ed. Paul Adams et al. (Minnesota: University of Minnesota Press, 2001), 403–25. Hereafter cited as *TPE*.
6. Tony Fry, "Design, Ethics and Identity," *Design Philosophy Papers* 3 (2006): 1–3.
7. Anne-Marie Willis, "Ontological Designing," *Design Philosophy Papers* 2 (2006): 1–11. Hereafter cited as *OD*.

8. Henri Bortoft, *The Wholeness of Nature: Goethe's Way of Science* (New York: Floris Books, 1996). Hereafter cited as *WN*.

9. Tim Robinson, *Stones of Aran: Pilgrimage* (London: Penguin, 1991).

10. J. E. Malpas, *Place and Experience* (Cambridge: Cambridge University Press, 1999). Hereafter cited as *PE*.

11. Laurene Vaughan, "Emplacing Local Invention" in *Studies in Material Thinking Journal* 1, 2008, accessed December 5, 2008, http://www.aut.ac.nz/material_thinking/materi-al-thinking2/currentissue.html.

12. We would like to acknowledge the project participants and co-authors of the project: Neal Haslem, Daphne Shao, Tania Ivanka, Lizzie Glickfeld, Nurul Rahman, Jeremy Yuille, Rebecca Nally, and Jess Atkinson. We also thank Miek Dunbar, whose input was instrumental in curating and developing the project outcome, and we acknowledge the support of RMIT University in the writing of this chapter.

Chapter Seven

Representing the City

Complementing Science and Technology with Art

William Cartwright

Understanding how technology works is important, but the partnership between art and science, and their contributions to the discipline, are as important. Art provides the "public face" of cartography (and if we include the cartographer's passion when designing particular products, perhaps the soul as well), and science complements this by ensuring that what is presented is scientifically correct and what could be called "scientifically elegant" as well.

INTRODUCTION

Science or technology, it is argued, need not always take on the primary roles in cartography. Technology is needed to ensure that the designed product can be produced and delivered, and science is necessary to ensure "correct" and rigorous products. However, the resulting artifact, designed and produced by balancing the art, science and technology attributes, as a street artist juggler might balance a chainsaw, a watermelon, and a table tennis ball, has recently been biased toward science and technology, with art being relegated to the position of afterthought (thinking about the art elements after the product's specifications are locked within a science foundation and technology-driven production and delivery "envelope"). Cartography is different from other contemporary disciplines insofar as it can design, develop, and deliver products with an art or a technology or a science "flavor." But we need to address how to make art-biased cartography as relevant as science or technology-biased cartography.

An example of this is using a number of artefacts to gain visual information about a city. It is argued: if different media are used, then a different viewpoint is provided, and perhaps a different interpretation of the city is had. The three following figures illustrate this. Figure 7.1 shows a landscape depicted as a painting; in figure 7.2 as planimetric map; and in figure 7.3 as a satellite image. A painting provides one representation of geography, the map another and the satellite image another. Compared to "standard" maps, like street directories and topographic maps, artistic interpretations can provide completely different viewpoints of geography. Users are provided with different platforms from which to view "reality."

With the current provision of conventional maps (including those delivered on mobile devices and the World Wide Web the question arises: are the users of conventional artefacts for depicting geographical information only being allowed to view information produced by one particular type of hard/sharp "pencil" drawing? Would there be a better understanding of the real world if there were many different methods for its depiction? It is argued

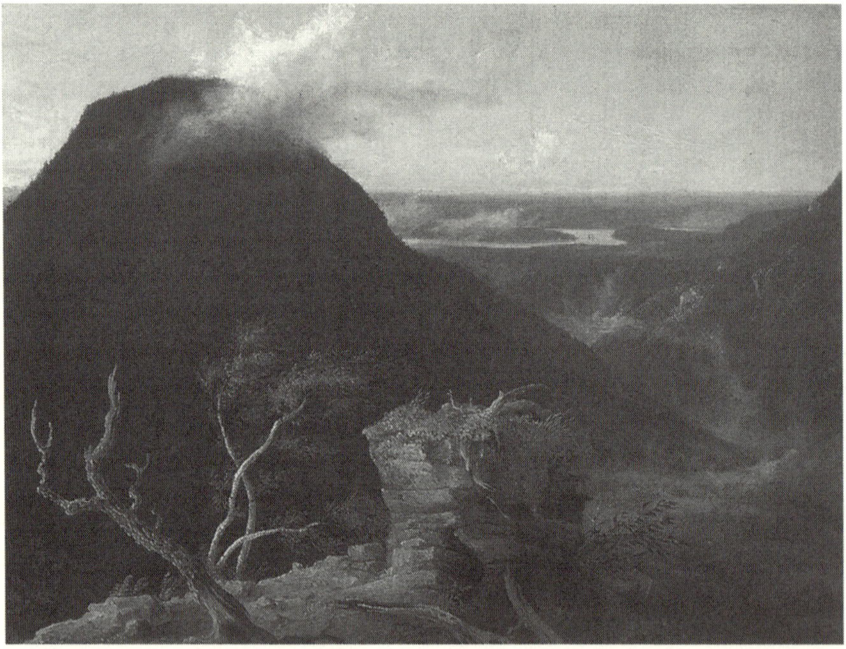

Figure 7.1. Sunny Morning on the Hudson River. Thomas Cole, 1827.1 This image is in the public domain because its copyright has expired. "Thomas Cole: Sunny Morning on the Hudson River" ***Outdoor Literature*****, last modified August 16, 2005. This image is in the public domain because its copyright has expired, http://www.isu.edu/~wattron/OLCole6.html.**

that the present devices for viewing the complex relationships that comprise the real world, even with the application of new technology, still only allow users to see that information in one manner: the hard, precise, sharp pencil manner that is a legacy of paper mapping. It is further argued that there is a need to investigate how the use of other "softer" presentation methods complementing today's generally "hard map" biased devices, portray geography and the mental images of constructed reality.

SCIENCE AND TECHNOLOGY: RESTRICTIONS IMPOSED ON CARTOGRAPHIC REPRESENTATIONS

Science and technology has provided cartography with the means for providing accurate maps in a timely and efficient manner. However, the loss of control over part of the design process is at a cost. In order to apply technology to map production and replication cartographers were required to amend their documents so that they accorded with the technology used to produce them or to communicate the geographical "message." This section of the discussion looks at some of the technologies adopted by cartography and how they affected the actual design of the map.

PRINTING

Aligned to paper map production, printers became part of the map production team and in many ways they dictated the "look" of maps for many years due to their technologically-imposed specifications on the map production process.

Map design had to follow function, and cartographers had to adapt their design and production to take into account the particular restrictions which printing placed on maps. Printing formalised map design and the actual look of maps reflected this. Take for example the map shown in figure 7.4. The printing process allowed for sharp representations of the city to be made and reproduced. These images, while faithful representations of the grid structure of the streets did not include any information relating to the human factors of the city.

However, it must be noted that many maps produced via the printing press do have a certain quality that can be termed "elegant cartography." Here, the precise replication of detail provides a map that, while still designed to conform to the demands of the printing press, can be admired as an excellent example of the cartographer's work and the engraver's skill. The map of Dublin in figure 7.5, while representing the city using the same printing method, does afford a greater understanding of it because detailed

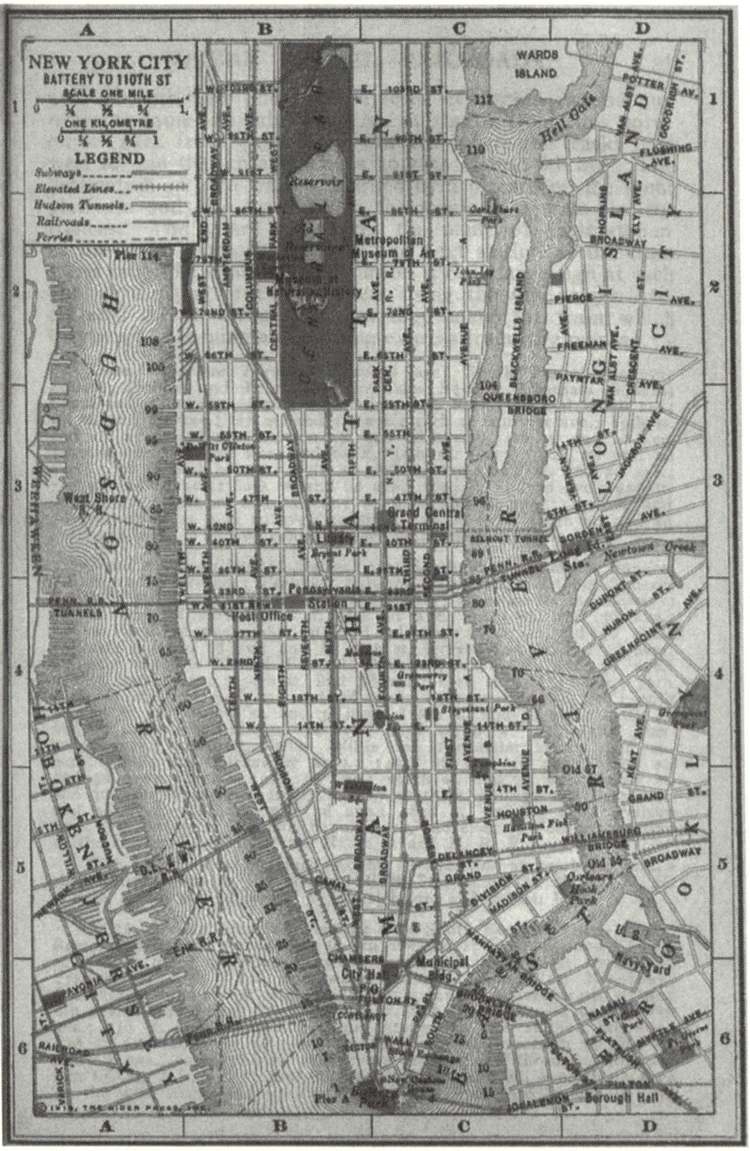

Figure 7.2. New York (Battery to 110th Street) 1916. "New York City Battery to 110th Street," Rider's New York City, Henry Holt and Company, 1916. Perry-Castañeda Library (PCL) Map Collection, last modified May 16, 2001. Courtesy of the University of Texas Libraries, the University of Texas at Austin. PCL Map Collection materials are in the public, are not copyrighted and no permission is needed to copy them, http://www.lib.utexas.edu/maps/historical/manhattan_1916.jpg.

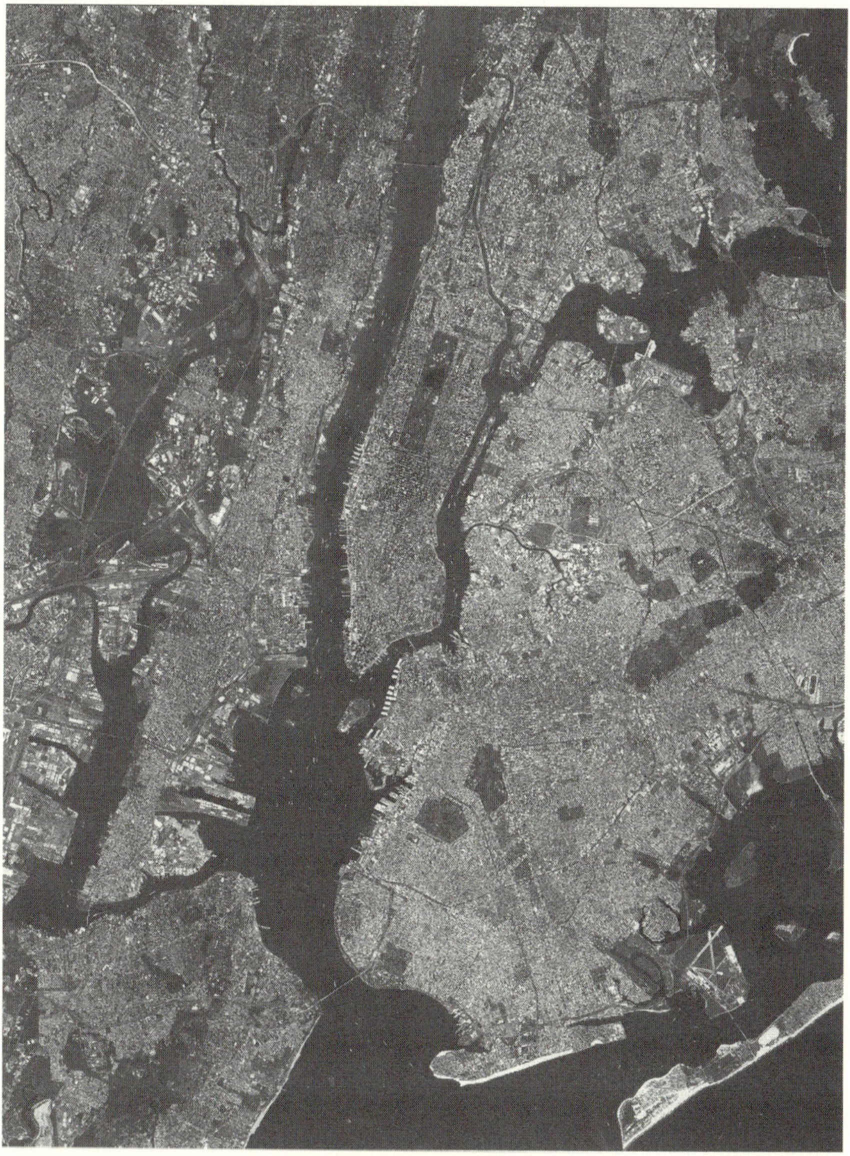

Figure 7.3. False-color satellite image of greater New York City (September 8, 2002). Courtesy of NASA, last modified September 23, 2008, at http://earthobservatory.nasa.gov/images/imagerecords/3000/3678/aster_newyorkcity_lrg.jpg. Most materials, including this image, published on the Earth Observatory at http://earthobservatory.nasa.gov/ImageUse/, including images, are freely available for re-publication or re-use, including commercial purposes.

three dimensional (3D) representation of individual buildings allows them to be identified and their function considered.

Calvino's *Invisible Cities* gives some insight into what an articulate reader may wish to read into maps:

> . . . relationships between the measurements of its space and the events of its past: the height of a lamp-post and the distance from the ground of a hanged usurper's swaying feet; the line strung from the lamp-post to the railing opposite and the festoons that decorate the course of the queen's nuptial procession; the height of that railing and the leap of the adulterer who climbed over it at dawn; the tilt of a guttering cat's progress along it as he slips into the same window; the firing range of a gunboat which has suddenly appeared beyond the cape and the bomb that destroys the guttering; the rips in the fishnet and the three old men seated on the dock mending nets and telling each other for the hundredth time the story of the gunboat of the usurper, who some say was the queen's illegitimate son, abandoned in his swaddling clothes there on the dock (*IC*, 10).

INTRODUCTION OF COMPUTERS

With the invention of the computer everything changed in the scientific world, including cartography. Cartography applied computer graphics for artwork production and output. Unfortunately many early, inferior maps produced with these early computer systems were readily accepted as substitutes for the precise and elegant scribed and printed alternatives because they were produced quickly and by new computer systems. Because the results of many calculations could be displayed using the newly adopted computer drawing packages, which drew on massive databases, users were sometimes willing to accept quickly produced, crude outputs.

Examples of these were American Standard Code for Information Interchange (ASCII) maps, which were simple maps with variations in greytone density achieved by overprinting characters on a black and white printer. The demands of this method of production made the map design accord with the replication technique, and the results were chunky and almost illegible when more detailed information was needed. As long as these maps were not exposed to too much critical analysis, all appeared to be well with the cartographics. But, still, the attempts to portray spatial information using early computer graphics systems did produce some below-standard maps. However, once mastered, this technology can also provide beautiful maps.

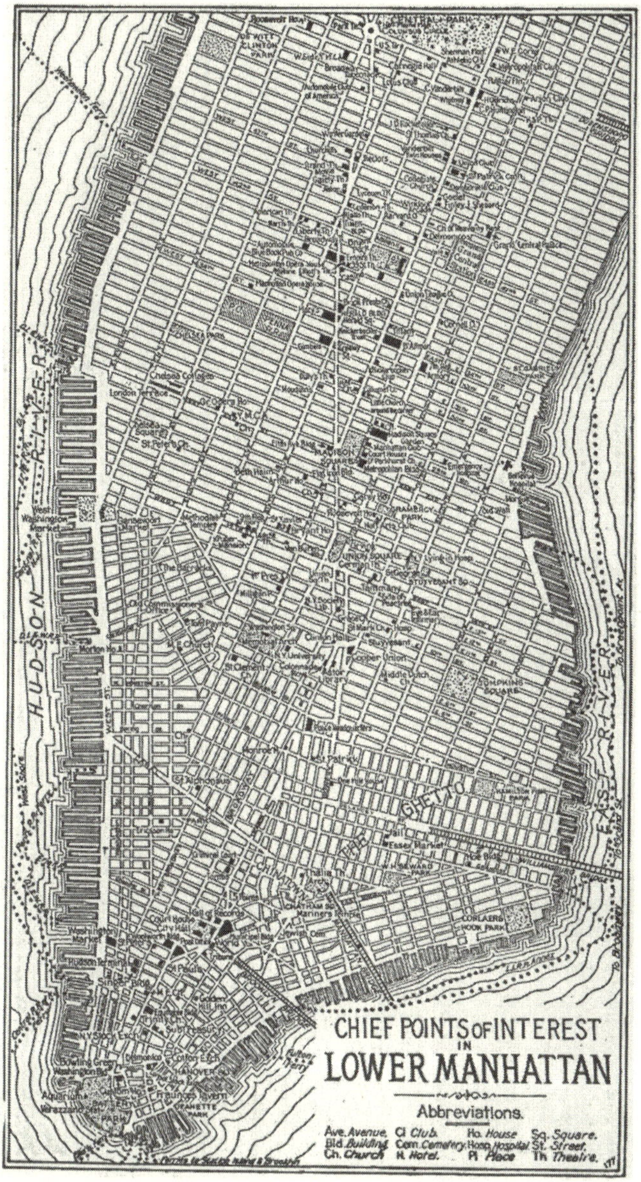

Figure 7.4. New York - Lower Manhattan. "Chief Points of Interest in Upper Manhattan" *Automobile Blue Book, 1920*. PCL Map Collection, last modified May 16, 2001. Courtesy of the University of Texas Libraries, the University of Texas at Austin. PCL Map Collection materials are in the public, are not copyrighted and no permission is needed to copy them, http://www.lib.utexas.edu/maps/histori-cal/manhattan_lower_1920.jpg.

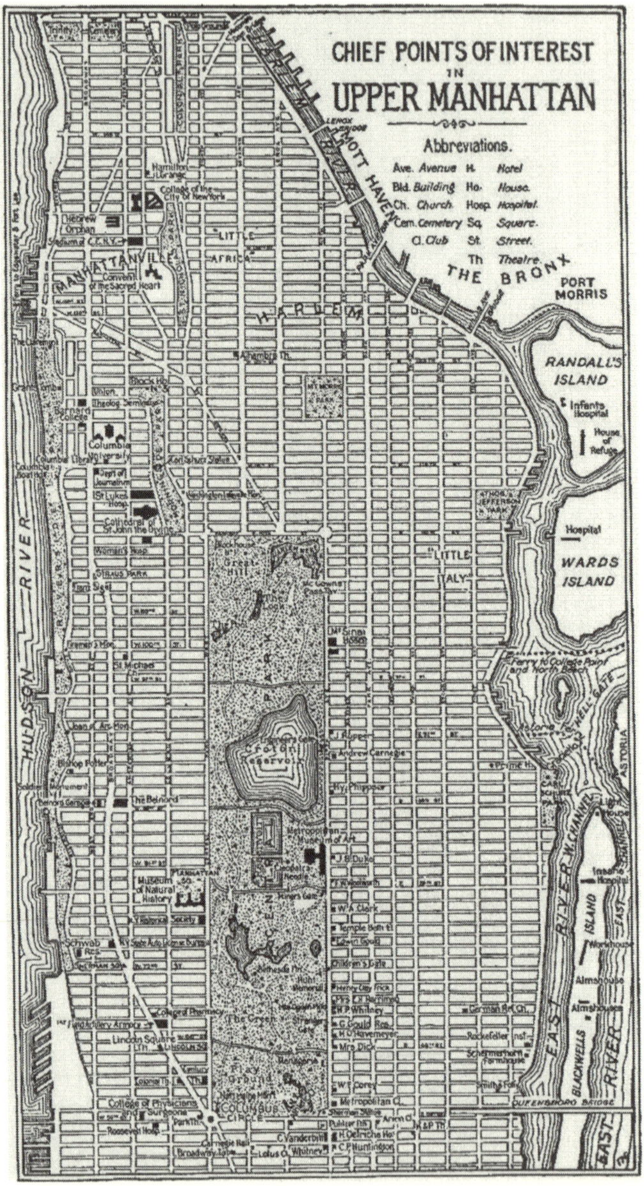

Figure 7.5. Upper Manhattan, 1920. "Chief Points of Interest in Upper Manhattan" *Automobile Blue Book, 1920*. PCL Map Collection, last modified May 16, 2001. Courtesy of the University of Texas Libraries, the University of Texas at Austin. PCL Map Collection materials are in the public, are not copyrighted and no permission is needed to copy them, http://www.lib.utexas.edu/maps/historical/manhattan_lower_1920.jpg.

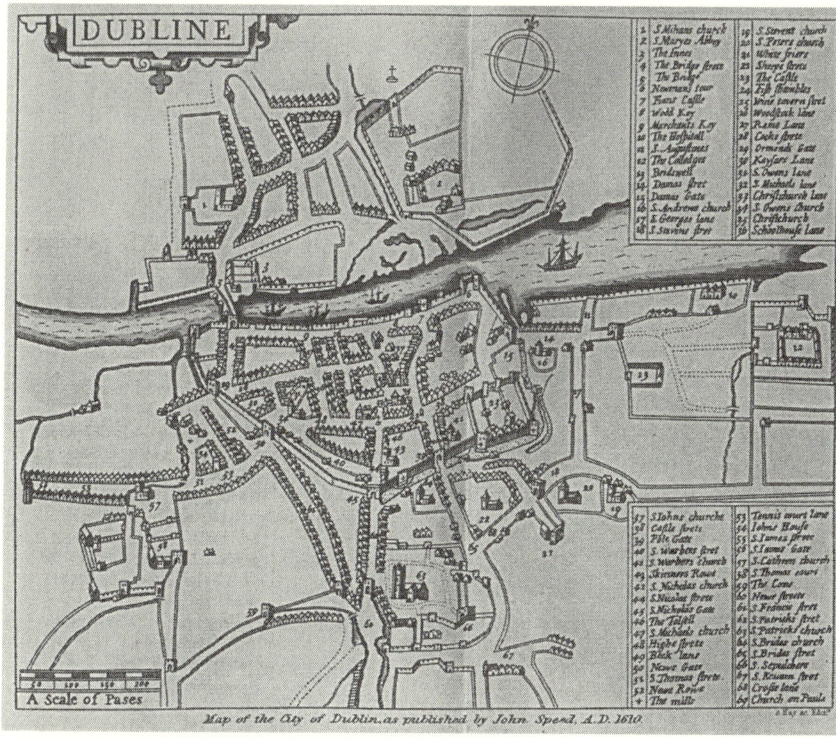

Figure 7.6. ***City of Dublin*, John Speed, 1610. Som Norsk by L. J. Vogt, H. Aschehoug and Co. 1896. PCL Map Collection, last modified May 16, 2001. Courtesy of the University of Texas Libraries, the University of Texas at Austin. PCL Map Collection materials are in the public, are not copyrighted and no permission is needed to copy them, http://www.lib.utexas.edu/maps/historical/dublin_1610_1896.jpg.**

DISCRETE MEDIA

The introduction of discrete media saw maps being produced on CD-ROM. Initially, the potential of the large storage capacity of CD-ROMs for the distribution of geographical information fostered interest in publishing digital maps using a new medium (*CCD*). Products like the Digital Chart of the World (DCW) and the World Vector Shoreline[1] were some of the first to exploit this storage medium. The first maps mimicked their paper cousins, but later on added interaction and user tools. British company, Nextbase produced a street directory of London, which included innovations like layers of information that could be turned on and off enabling information to be made available when required and enhancing what could be immediately viewed—an apparent electronic version of a paper street directory. What was produced was still a "standard" street map with interactivity added. The media was not fully exploited and the maps were only electronic reproductions of paper ones.

Later, the appearance of Desktop Publishing (DTP) packages made every graphic artist a cartographer. A flood of "crude" (from a design perspective) computer-generated maps depicted everything about everywhere. This relatively recent transition of mapping from large electronic purpose-built systems housed in a map production company to individual desktops has also been a revolution for small map producers. Individual cartographers were able to produce maps as sophisticated as their corporate counterparts. Equipped with a powerful microcomputer plus a scanner, plotter/printer, and modem the individual becomes part of the distributed digital electronic mapping community. Figure 7.6 illustrates how DTP can be used to illustrate the city, over and underground.

THE WEB

The arrival of Web publishing initially mimicked somewhat the maps produced in the early applications of CD-ROM. Paper maps were scanned and collections provided on this optical media. Maps were not designed specifically for the medium and compromise products resulted. Early Web mapping sites provided access to map collections. These were initially scanned maps, like those in the CIA World Fact Book[2] and the PCL Map Collection.[3] While delivered immediately, these maps provided low-resolution replications of maps that were printed on paper in higher resolution than the 72 dpi (dots per inch) computer screens on which they were displayed. Even later, the maps provided by this media were woeful in terms of design.

While generated quickly by computer and delivered by the Web, they did not advance cartographic aesthetics at all. Later, as bandwidth increased and map designers addressed the Web as a real publishing media map design improved. They were designed to work with the restrictions imposed by the communications medium. This can be seen to parallel somewhat the printer-cartographer relationship that evolved with printed maps. Map design and production methods were altered to fall into place with Web publishing. Maps like the interactive Virtual Reality Modeling (VRML) 3D application of Sydney Road, Brunswick (figure 7.7) illustrates an example of a product that worked with the foibles of the Web.

Most recently maps have been published on the Web by producer-users using a process called "mash-ups" with Web 2.0 and Social Software. Individuals and groups of individuals use the Web to provide and share information, including geographical information. It provides a new model for collab-

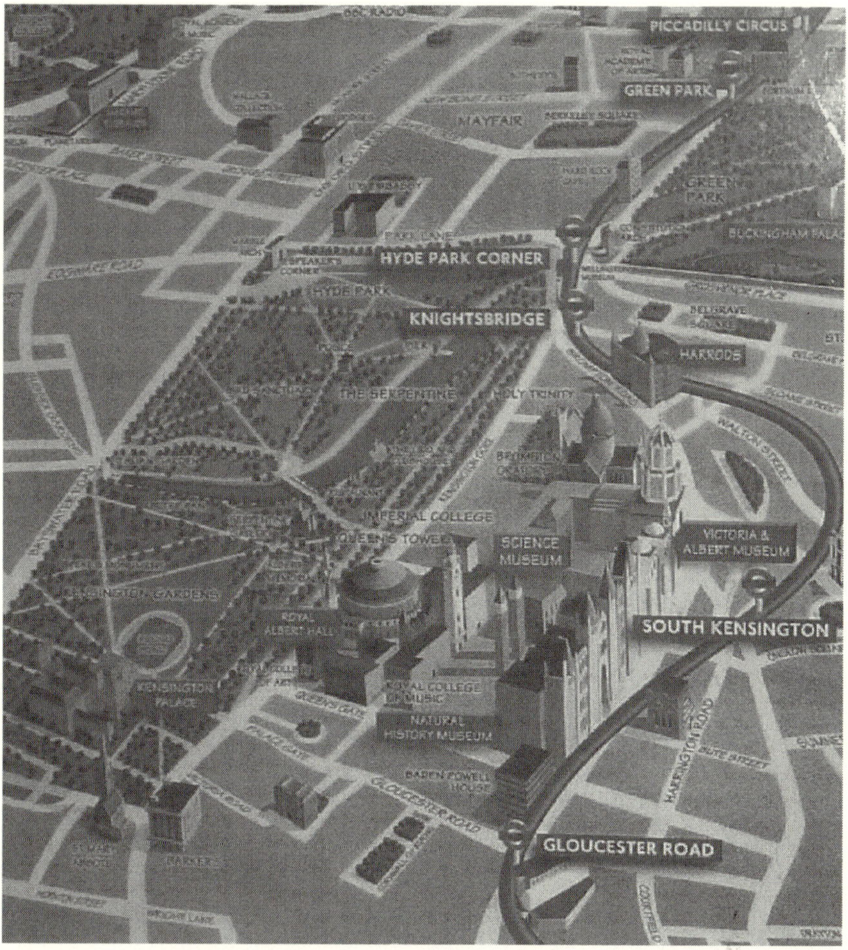

Figure 7.7. Image of London Underground / above ground poster. Poster photographed on a London Underground train, 2007. Photograph: William Cartwright 2007. Courtesy of William Cartwright.

orating and publishing. An example of this type of product, developed as part of a research project by the author and produced and delivered using *Google Earth* is shown in figure 7.8.

THE CASE FOR MAPS

Maps show one "view" of geography—that composed and provided by cartographers. However, according to Peters, a postmodern account of geography would not consider any one representation of a place as any more valid than any other representation (*CLA*). There are numerous contemporary landscape artists who, according to Peters " . . . represent an infinite subject in a finite space" (*CLA*, 1). For example Benjamin Edwards is exploring the "architecture of suburbia" and developing pieces that depict landscapes that are created by the American Interstate highway (*CLA*, 2).

However, it has been argued by geographer/cartographers like Castner that the map model works best, and that users have to appreciate the "grammar" of cartography in order to fully understand the "language" of maps and how it depicts geography, including the associated "lies" that maps have to tell to illustrate the "truth" about what the map reader needs to see on the map in order to have the best view of reality. The defined geographical "picture" that is used by the tour or itinerary provides only a small view of reality and therefore precludes a true appreciation of what constitutes the "real world," and where the user "fits" into that world.[4]

Considering that cartography has been described in terms of science, art and technology, it is perhaps necessary to re-visit the description in the light of the application of computer technologies. Is cartography any different when delivered using computer-driven devices? Does cartography need to be re-defined because of the revolution that has taken place with both the way in which information is now communicated and the type of information that can be transferred, almost instantaneously, globally?

An argument could be put that cartographers become involved in the elements of cartography that they have both mastered (either academically or technically, or both) and that they also enjoy doing. Personal satisfaction in producing an elegant and aesthetically pleasing design or mastering some scientific problem can be a major part of what motivates cartographers and encourages them to refine further the skills and scientific strategies that are unique to cartography. However, this could lead to a situation where they could be uninterested in producing maps for anyone but themselves.

In the current situation, where maps are no longer *the* focus for every interpretation of geography, should cartographers view what they do differently, and do consumers of cartographic products influence the art/science/technology balance? Contemporary cartography could be seen to have as much to do with making a movie as producing a scientific document. While the scientific integrity of all cartographic products is as important now as ever (perhaps more important because of the casual appearance given by the immediate façade of contemporary mapping) the art components need to be considered as equal partners to their scientific counterparts. Indeed science and technology may only be getting in the way of best depicting geography.

CONCLUSION

Understanding science and harnessing technology are important to cartography, but the partnership between art, science and technology is equally important. In my opinion, art provides the "public face" of cartography (and the cartographer's passion when designing particular products perhaps the soul) and science complements this by ensuring that what is presented is scientifically correct, and what could be called "scientifically elegant" as well. As recent developments of the Web show, science and technology are not the only means of depicting geography.

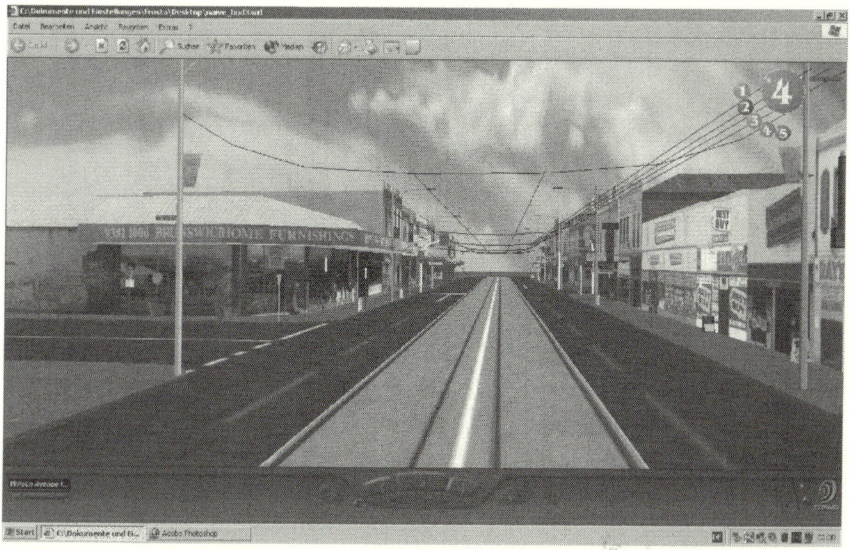

Figure 7.8. Web: VRML "world" of Sydney Road, Brunswick (Australia). Courtesy of William Cartwright.

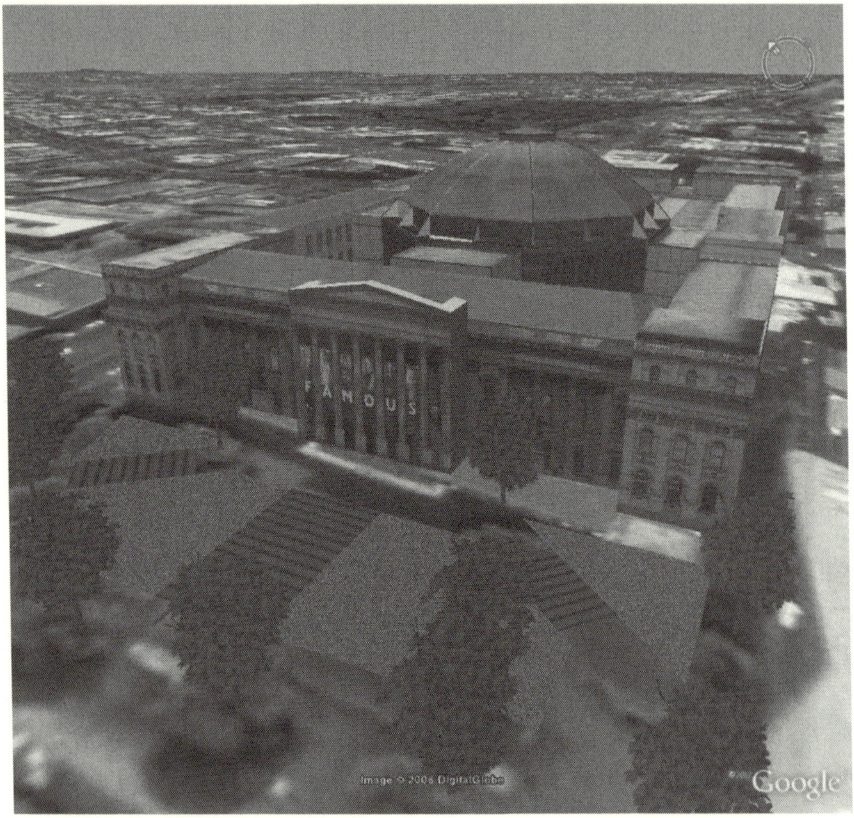

Figure 7.9. Work in progress, William Cartwright.

NOTES

1. B. J. Lauer, "Mapping Information on CD-ROM," Technical papers of the *1991 ACSM ASPRS Annual Convention*. Baltimore: ACSM-ASPRS 2 (1991): 187–93.

2. https://www.cia.gov/cia/publications/factbook/docs/refmaps.html.

3. http://www.lib.utexas.edu/maps.

4. This discussion comes from William Cartwright, "Art and Cartographic Communication." In *Art and Cartography*, W. Cartwright, G. Gartner, and A. Lehn (eds.) (Berlin Heidelberg: Springer-Verlag), 19.

Chapter Eight

Embodied Encounters

The Photographic Seeds of Rolf de Heer and Peter Djigirr's Ten Canoes

Linda Daley

Photographic technologies are usually viewed in postcolonial discussions as being exploitative of indigenous peoples and themes. In the context of anthropological photography, this perspective reaches its high point in being viewed as the West's privileging of vision as the basis for all knowledge of the Other. Christopher Pinney says the familiar, panoptic view of the twinned history of anthropology and photography accounts for only one of its two alternative histories.[1] Its other history is marked by "moments of unease," one that is fractured by the competing interpretive claims of realism on the one hand and expressionism on the other (*AP*, 72).

POSSIBLE FUTURES OF ANTHROPOLOGICAL PHOTOGRAPHY

Just as Pinney reflects on two alternative histories of anthropological photography we can also conceive of its other possible futures. My aim is not to dismiss the relevance of postcolonial critiques of anthropological photography, but rather to consider the enabling possibilities of camera technologies and their associated practices through analyses not dominated by a politics of the gaze or by theories of representation that focus exclusively on the image as the source of visual meaning. Such frameworks tend to view the image as fixed or static in meaning across time and contexts, and they overshadow other ways of analyzing relations between the camera, culture(s), and ways of seeing that are more attuned to contemporary indigenous experiences.

Approaches that are variously named "looking relations" and "seeing and seen-ness" reflect the changed and changing nature of indigenous subjects' relations to camera technologies, and also the changed and changing nature of relations between museums and the source communities where the photographs originated.[2] These latter approaches also emphasize the material dimensions of photographs to sense making in considering the embodied encounter by viewers of the physical substrate of the image as well as its presentational form.

In analyzing photographs more broadly than their image content alone, and by investing them with a social biography acquired through the various stages and modes of encounter in their consumption clearly favors historical print photographs. It also invites the question that Pinney asks in his more recent work: what is the relationship between photographs and their contexts that makes some more capable of recoding than others, which are more resistant, and others still, completely intractable? (*POH*, 4) My focus here is to respond to Pinney's question with reference to Donald Thomson's ethnographic photographs and their more recent role in the production of Australia's "most expensive and ambitious" intercultural feature film, *Ten Canoes*[3]—the first Australian feature spoken in an indigenous language.[4]

Thomson took more than 2,500 black-and-white photographs of sacred and everyday activities, and individual and group portraits of Yolgnu, during his two field expeditions of 1935–1937 and 1942–1943 to Arnhem Land in the Northern Territory of Australia. Following the photographs' repatriation in the early 1990s to their source community of Ramingining, a small number of the images have formed the narrative seed of the drama and are re-staged within the story. By "re-staged," I mean a small number are performed or mimicked by actors who are descendants of the subjects in the original Thomson photographs. It is the nature of the casting for the film, and in particular the miming of the original images by the actors, that is a highly ingenious means of appropriating the images as well affording a means of cultural renewal for the people of Ramingining. My claim is that the Thomson images have been recoded as records of scientific evidence because of the nature of relations between the photographer and the photographed and the practices and conditions under which the images were produced. These conditions and practices inflected the images with the possibility of their recoding at a future time and in other contexts to those of their production. In order to account for this claim, I need to place the photographs within their social biography and to do so from their most recent incarnation within the feature film.

YOLGNU AND WESTERN MIMETIC PRACTICES INTERTWINE IN *TEN CANOES*

Five images have been restaged within the central third of the ninety-two minute feature as stilled rather than frozen images around which the drama is built. The narrative has two main time frames: the first is of the framing story which is filmed in black-and-white, uses a fixed camera, and according to the film's press kit, is set "about a thousand years ago."[5] The frame story minimally dramatises the central Thomson photograph of ten men in the canoes from which the film takes its title. It tells the story from the perspective of one of the men, Minygululu, who learns that his younger brother, Dayindi, is in love with Minygululu's third and youngest wife. The older brother tells Dayindi a parable from mythical time, from the Dreaming, as a lesson in the importance of living the proper way according to the law. The parable from the Dreaming opens to the second time period of the narrative, and is filmed in color. In contrast to the frame story, this narrative uses a mobile camera and involves a more complex plot and characterization. It is a parable that is long in the telling, taking all the time of the canoe-making, the swamp travelling, and the goose-egg gathering to impart. The two time frames—one of history and the other of myth—intertwine throughout the duration of the story. Internationally renowned actor and traditional land owner, David Gulpilil's voice-over provides an outer frame to these two time periods, and therefore technically offers a third time period, that of the contemporary present.

Defying conventional film practices that would have actors selected according to talent and appearance, the untrained actors in the feature were cast according to a strict kinship correspondence between the individuals in the relevant Thomson photographs and their descendants. The result is a creative mix of realism and expressionism that goes to the core of Yolgnu notions of mimesis and representation, and demonstrates the fracture in anthropological photography of which Pinney speaks. Visual anthropologist Jennifer Deger explains this aspect of Yolgnu cosmology in terms of a neo-Heideggerian idea of "presencing" that "challenge[s] Western conceptions of representational theory" (*SSM*, 99). She accounts for the "presencing" that occurs with photography by explaining that recording technologies like the camera amplify a mode of relationship between a subject and viewer of the visual phenomenon that is central to all Yolgnu mimetic practices such as dance, ceremony, and art. The link is not simply between the subject and viewer—in this case, of the photograph—but also between the Ancestor with whom the subject of the photograph is linked and that of the viewer. Like all other mimetic practices—although amplified because of its reproducibility—the Thomson photographs link the Yolgnu viewer to an entire web of relations in the Ancestral realm.

Non-indigenous viewers—and perhaps non-Yolgnu indigenous viewers too—see in the central Thomson photograph ten men in canoes on a swamp, whereas Yolgnu viewers see ten men as well as the invisible Ancestral elements with which those Yolgnu are ontologically connected and presenced through those ten empirical beings. On this basis, who acts in the part of a character in a dramatised re-staging of the photographs has significant implications not only because of the genealogical link to the empirical person in the photograph, but also because of the mythical beings to which the photographic subject was linked, and gives presence, in the photograph.

Thus, the kinship structure Yolgnu law demands for its social order is transposed to the casting for the filmed dramatic performance. From the perspective of Yolgnu cosmology, the restaging of the images in the film enables mythical beings as well as empirical beings to be presenced through the body of a present day actor whose acting is faithful to both Yolgnu cosmology and the visual representation of that cosmology as documented extensively by Thomson. Miming the compositions of Thomson's images in the film is an active form of re-presencing as well as a representation of that re-presencing in the photographs through the bodily act of performance by each of the actors. The displacement of Thomson's photographs as ethnographic records occurs not simply because their image content is presented in a different format (that of film) and through a different context (that of narrative cinema), but also because of the appropriation of the photographs' image content by the actors whose bodies act as conduits of the images' display-as-displacement.

RECIPROCAL ENGAGEMENT WITH THE PHOTOGRAPHIC IMAGES

It is worthy of this ingenious form of engagement with the Thomson images to reflect on the challenge to Western assumptions about seeing and being and sense-making this kind of engagement undertakes. Consistent with these Western assumptions, the sense of touch, for example, is undervalued in much photography analysis. However, the experience of Yolgnu with the Thomson photographic archive demonstrates its importance.

In the documentary about making the feature, *The Balanda and the Bark Canoes* (dir. Molly Reynolds, Tania Nehme, Rolf de Heer, 2006) we can see multiple, folder-bound photocopies of the photographs being closely scrutinized both privately and collectively on the film set: fingers tracing along the lines of the canoe while pointing out detail about its prow; cast sitting with directors around the fallen paperbark gums about to be constructed as props for the film discussing detail in the photographs and diagrams Thomson made. In their photocopied and laminated formats for the film set, the photographs are reformatted to allow for a greater physical engagement with the originals by the community and film cast, which is, arguably, only the prelude to the more intense physical engagement with the images in their bodily appropriation by particular actors from the community.

Susan Stewart's phenomenological descriptions of memory and materiality remind us that the eye is located within a larger corporeal context of sense perception and understanding that includes touch as well as smell and taste. These ambient senses link to what is perceived and perceptible to memory formation even though they are conventionally subordinated to sight as the sense that is thought to most readily organize the visual field (*MMD*, 31–32). Stewart claims that even so-called properly visual objects such as holograms, rainbows, and dream images, entail physical processes in that being in contact with one of these objects means to be moved by it and to have the pressure of its existence brought into relation with the pressure of our own body. She further claims that whereas there is a temporal immediacy involved in visual perception that has its parallel of spatial immediacy in tactile perception, it is the latter rather than the former that entails movement and change on our part: we turn the object, and in turn, it turns us, and in that turning, it takes time, and in this stretch of time there is a connecting link between the subject and object of visual/tactile attention.

The encounters by present-day Yolgnu of Thomson's photographs of their forebears that trace a line of generations stretching for millennia, invite the touching, pointing, and stroking of them as much because the memory formation is too great to be left to the eye alone, but also because the photographs as image-objects invite that touching as a form of reciprocal connection. Frances Djulibing, who plays Nowlingu in the Dreaming part of the narrative, saw a Thomson photograph of her grandmother, Yilpa, wearing a breast harness that inspired her to make and wear one as part of her body gear in the film.[6] Just as these images of sacred and everyday activities and of individual and group portraits work on the descendants of Thomson's subjects, so too do the viewers work on Thomson's photographs in their restaging in the feature film. This gesture entails the incorporation of the past time that the Thomson image signifies, and also the event of the original encounter between the photographer and the photographed through the bodies of the Yolgnu actors. Quite literally, that re-connection with the past is occurring through the bodies of the actors in the film through their encounters with Thomson's photographs.

YOLGNU COLLABORATIONS DURING "THOMSON TIME"

Thomson worked in three frontier regions during his professional life: Cape York (Queensland), Arnhem Land (Northern Territory), and the Great Sandy

Desert and Gibson Desert (Western Australia). It was only during his time in Arnhem Land that he built dark rooms in the field, thereby enabling him to develop the plates and show the photographs to his subjects (*DTM*, 57). There is, however, no evidence that Thomson gave any of the prints to his subjects.[7] During the period known by locals as "Thomson time," the photographs were produced exclusively as ethnographic records of a way of life that he believed would soon be extinct due to the hazards of European contact on indigenous communities.[8] Anthropologist Athol Chase says "Thomson time" is the phrase used approvingly by locals to refer to a recent period of history—still remembered by some of the locals—when Thomson worked with "the last of the 'bush people'" whose knowledge and expertise was fully traditional and little affected by the contact process" (*AH*, 109–10). While Thomson's belief in the extinction of a traditional way of life commonly prevailed in both popular as well as anthropological discourses, in Thomson's case it was tempered by his methods and approach to his fieldwork and the nature of his relations with the people with whom he worked, which, as anthropologist Peter Sutton claims, is reflected in Thomson's photography (*DTM*, 154). Yolgnu elder, Gatjil Djerkurra describes the Thomson photographic archive as "the best pictorial record of our culture and old people that can now be produced" (*DTA*, vii).

A highly skilled photographer from a young age, Thomson adopted the glass plate camera technology in the field for its high tonal quality, and viewed photography as a field method in its own right rather than a mere supplement to the written word (*DTA*, xiii). He had a meticulous approach to identifying, classifying and documenting not only his photography, but also his field notes and drawings. For example, of the 7,500 material culture items he collected in his life-time, he not only attached detailed labels that included the date of collection and details about the object recorded in the local language, he also frequently photographed the object in use before its collection (*DTM*, 3). The camera technology he used was cumbersome, requiring considerable care and preparation in the loading, unloading, and reloading of the fragile plates in the slide holders, and often in complete darkness when the water temperature allowed development (*DTM*, 47). Many of his photographs were staged for either logistical or compositional reasons; for example, to take advantage of a certain light, environmental condition, or ceremonial event, and he would often request that his subjects remove their clothing before the picture was taken. Thomson believed that all visible signs of European contact degraded and diminished the appearance of Aboriginal people. It is clear from his photography that it required not only a considerable amount of preparation and skill on his part, but also the collaboration of his subjects (*DTM*, 57).

Thomson combined the rigor of his undergraduate training in the natural sciences with a participant observation approach to his fieldwork—somewhat ahead of his time and out of step with his anthropology training—in gathering thick description of many facets of sacred and everyday life. He recorded information from Yolgnu in their own dialects, which he learned in the field and would then transcribe phonetically (*DTA*, 18–19).

Although Chase describes Thomson time as a period "little affected by the contact process," it was also a period on the threshold of what would be felt by indigenous peoples as rapid and traumatic change brought about by the corrosive effects of colonialism. Thomson's arrival in Arnhem Land in 1935 was precipitated by his self-initiated mission to assist Yolgnu over the miscarriage of justice against Aboriginal men who had killed Japanese crewmen on luggers fishing for trepang (*bêche-de-mer*) in the bays of Arnhem Land in defense of their territory and with the prospect of total depletion of their food source. Newspaper reports of killings over a number of years had started to attain mythic proportions Australia-wide, and suggestions of punitive expeditions were being made (*DTA*, 6). Thomson sought and eventually gained permission from government authorities to undertake his mission. His duties were to establish friendly relations with the Aborigines of Arnhem Land; communicate to the Aborigines the gravity of acts of murder and robbery of both fellow Aborigines as well as Europeans; to report cases of serious illnesses; to study and report on the language, ceremonies and customs of the clan groups (*DTA*, 32). Thomson believed this kind of mission could only be based on science rather than either the moralism of the church, or the force of the law as carried out in the preceding decades by the police in the Northern Territory. Thomson says: "I realised that here was an opportunity which would not occur again in a lifetime, to demonstrate the practical value of an anthropological approach, to avert disaster for the Aboriginal people of eastern Arnhem Land and to pave the way for a completely new policy in administration" (*DTA*, 25).

Upon his first arrival in Calendon Bay in Arnhem Land, Thomson engaged with two senior Yolgnu men, Raiwalla and Wonggu, and explained his mission. He "sat down" with Wonggu for a number of days, having brought message sticks from Wonggu's sons in a Darwin gaol, with the result of these talks being a message stick from Wonggu given to Thomson, which he intended Thomson to carry to the authorities in the "Big Country." The message stick visualises Wonggu as a respected lawman who could maintain order among his group (*DTA*, 80). "Thomson time" is also remembered by present day Yolgnu as a period of close, intensive efforts at intercultural understanding, and Thomson's record-taking in the broadest sense, not only his photography, must be seen in this context of mutual respect and trust by Yolgnu who gave Thomson access to their life-world. The trust that underscored Thomson's time in Arnhem Land is probably most readily demonstrated by the fact that he was able to photograph the women and girls on Groote Eylandt, and make a study of their food gathering and domestic economy without any trouble and after only a brief prior visit. By contrast, according to Thomson, church missionaries who established a settlement on the island in 1921 did not see an Aboriginal woman for the first four years; and bird collector, W. McLennan, who spent months on the island in 1923, never saw an Aboriginal woman at all during his time there (*DTA*, 110).

RECODING OF PHOTOGRAPHS DEPICTING A "DYING RACE"

What is the nature of the recoding of Thomson's photographs by Yolgnu in the feature film? It is not the erasure of what Roland Barthes would call their "evidential force," understood as the photograph's unique capacity to depict the "there-then" as the "here-now" for its viewer (*CL*, 87). Arguably, this reality effect of photographs is precisely why they are valued by Arnhem Landers. Rather, it is their recoding of the "evidential force" of a "dying race" and as a "way of life that was soon to be extinct." Thomson's records were generated within an epistemological context and methodological process premised on the belief that traditional indigenous existence would not continue. Furthermore, ethnographic collections were built on the premise that the peoples whose material heritage was being collected were dying out, and that remnants of their culture should be preserved for the benefit of future generations. The notion that future generations should be the descendants of those with whom early collectors interacted was not considered at the time the collections were assembled by ethnographers like Thomson (*MSC*, 1). For all of Thomson's sensitivity to Yolgnu culture and practices, and his admiration for their ingenuity and intelligence, it is likely that Thomson held a similar view of the future of the indigenous groups with whom he worked.

Not only did the making of *Ten Canoes* enable an ingenious and enduring form of memory revival through the bodily appropriation of the Thomson images, it has also been a means of more tangible benefits such as a source of funding to re-learn the traditional skills of canoe making, fibre work and shelter construction for the community. The legal arrangement under which it was produced has become a model for film and other arts producers on intercultural projects in challenging some standard filmmaking practices, such as I have already mentioned with the casting, and also in regard to the property rights of the film sets, body wear, and props, which have been retained by the community of Ramingining rather than the film's producers (*TCS*, 5–14).

However, *Ten Canoes* has not been without criticism by indigenous and non-indigenous alike. While the film achieves the re-appropriation of the visual elements of the Thomson photographs, and goes some way to appropriating the time of Thomson through the production process for the cast, the representation of "Thomson time" in the film as a whole is missing. By visually quoting Thomson, but narratorially eliding the time of history to which Yolgnu attach his name, Yolgnu are largely represented in the film as outside history. This is not so much a loss for Thomson as for the people of Ramingining. The narrative frame of the period "a thousand years ago" that is visually reinforced through the use of the static camera, filmed in black-and-white and with minimal dramatisation of the period effectively consigns Yolgnu to a period outside of history reminiscent of the notion of pre-history that indigenous peoples have long fought against. The period of "a thousand years ago" not only pre-dates the European contact that "Thomson time" marks, it also pre-dates centuries of exchange between Yolgnu and the Maccassan peoples of Indonesia.

Studies by visual anthropologist, Gaynor Macdonald of Australian indigenous people's experiences of photographs demonstrate that when the colonial experience has been predominantly one of loss and denial of both myth and history, the re-integrative possibilities for indigenous peoples through photography is crucial. She argues that photographs act as markers for indigenous people to see themselves as "historical people" in that they "validate a history of engagement, of involvement, [as well as] of ancestry."[9]

Leaving aside criticism of the film's representational qualities in order to consider film's capacities for sense-making through the broader interpretive framework of camera technologies and their practices, we can see that *Ten Canoes* has enabled cultural renewal for the people of Ramingining. This settlement is the township nearest the Arafura Swamp where the film is largely set, with a shifting population of about 700 from several Yolgnu clans and three main language groups. The town did not exist when the Thomson photographs were taken, having emerged from the homelands movement in the 1960s when indigenous people in church and state missions were resettled on their country. Like many remote communities in Australia, Ramingining continues to suffer from economic and cultural marginalization in the decades since Thomson time. Frances Djulibing who mimes one of the Thomson images describes the importance of the film's production to her community: "Everything is changing, everything is going, going, gone now. The only thing the children know is some ceremony. . . . Maybe they gonna keep this film with them so they can put it in their head" (*TCK*). Similarly, Michael Dawu, who plays the part of one of the ten canoeists in the central Thomson image, says: "Aboriginal people . . . they was forgetting culture because every time we sit in town—sugar, damper, air-conditioner, light—we forget. We forget it long time . . . but for your memory, you have to go back, but my memory was gone" (*TCK*). It was David Gulpilil who brought the Thomson photographs to the attention of the film's producer and co-director, Rolf de Heer, with the intention of allowing "the people in the community and around the world to know how our ancestors lived, and to understand them."[10] By means of their display and displacement in the feature, and their role in the process of the film's production, Thomson's ethnographic records enable the film's participants—comprising most of the community of Ramingining—a means of integrating the past within the present and of mapping their remote community within a global geography through the film's screening in fifteen countries across three continents.

THE FUTURE-PAST OF ANTHROPOLOGICAL PHOTOGRAPHS

The Thomson images have changed format during their social biography from glass plates through to photocopied and laminated reproductions of prints, and more recently through their miming in the film. In their filmic format, the images have been brought out of their frozen and fragmented status in the archive to be returned to the flow of history as it were, to be re-presented in the film through the actors' bodies.

Two aspects of Thomson's Arnhem Land photographs invite speculation about the sense of future this meticulous recorder and inveterate collector of material culture had regarding what came to be a major component of his posthumously named Donald Thomson Collection. Unlike many homecomings of photographs to source communities, the repatriation of Thomson's Arnhem Land photographs is different in two respects. First is that for the first four decades of the photographs' existence, including the Collection as a whole, they were held by Thomson for his own personal reference with very few people knowing of their quantity or quality until after his death in 1970.[11] Thomson's family transferred the majority of the Collection (material culture items, field-notes, and photographs) to the University of Melbourne some years later, most of which was subsequently transferred to Museum Victoria. Apart from a few published within articles in scholarly anthropological journals, the Arnhem Land photographs were mostly seen only by the photographer and the photographed, and remained largely unseen by the anthropological community and the wider public for nearly four decades. When the photographs emerged from their relative obscurity, they did so at a time when anthropological and museological assumptions and practices were in transition from those that prevailed when Thomson was in Arnhem Land. At the moment of their coming out of Thomson's personal archive, his photographs were soon to be coming home, therein diminishing the photographs' institutional coding.

Second, while Thomson's general field practice was to name individual subjects in his portrait shots, the all-important image of the ten canoeists that captured Gulpilil's and de Heer's attention was without any identification until Museum Victoria senior indigenous curator, Lindy Allen undertook a photo elicitation project and consulted with Djimba elder, Tom Djumpurrpurr, and Ganalbingu elder, George Milpurrurru to identify the ten individuals and their clan groups in the image (*IAM*). To what extent this exchange between Allen and local elders would have provoked a wider community discussion about this particular photograph and its relation to the entire set of photographs returned to Ramingining, and since housed in the Bula Bula Arts Centre in 1993–1994, can only be surmised. But it seems plausible that the very absence of names, and therefore of the need to fill that absence with closer viewing, handling, and discussion could very well have provoked a greater interest in this image over many of the others where subjects were identified by Thomson. It is plausible that this very absence of naming further required the oral back-story that all photography engenders, and which Roslyn Poignant has described as photography's capacity to "seed a number of narratives."[12] It was this single image of the ten canoeists that Gulpilil showed de Heer during filming of *The Tracker* (2002) (which de Heer directed and in which Gulpilil had a lead role) with the plea to de Heer to make a film about his people. In a television interview, Gulpilil says of the motivation for the making of *Ten Canoes*: "I wanted to introduce Donald Thomson; was a true story of Dr Thomson. He met the traditional people and he recorded, and it was my uncles, my father and grandfather, and this is a story I wanted to come out" (*AJA*, 123–26).

Thomson's Arnhem Land images are recoded through the bodies of the Yolgnu actors as, arguably, the most intensive form of affective encounter possible with a photographic image. Through their entire bodily engagement with the images—from physically handling, touching, stroking, and tracing the many reformatted images from the Collection to the bodily mime of a small number by a few of the actors—Yolgnu have undertaken a process of incorporating the past of not only the skills, practices, and knowledge of their forebears' more traditional way of life, but also of that singular historical encounter of intercultural understanding between Yolgnu and Balanda (European) during "Thomson time." That process of recoding relies precisely on the "evidential force" of the photographic record, and the particular conditions of their production by Thomson in collaboration with his Arnhem Lander subjects.

NOTES

1. Christopher Pinney, "The Parallel Histories of Anthropology and Photography," in *Anthropology and Photography 1860–1920*, ed. Elizabeth Edwards (New Haven, CT: Yale University Press, 1992). Hereafter cited as *AP*.

2. See Faye D. Ginsburg, "Screen Memories. Resignifying the Traditional Indigenous Media," in *Media Worlds: Anthropology on a New Terrain*, ed. Faye D. Ginsburg, Lila Abu Lughod and Brian Larkin (Berkeley: University of California Press, 2002), 50; Christopher Jenks, "The Centrality of the Eye in Western Culture" in *Visual Culture*, ed. Christopher Jenks (London: Routledge, 1995), 12; and *Museum and Source Communities: A Routledge Reader*,

ed. Laura Peers and Alison K. Brown (London: Routledge, 2003), 1–16. Hereafter cited as *MSC*.

3. Rolf de Heer and Peter Dijgirr (dirs.), *Ten Canoes*, Palace Films. Australia, 2006.

4. Therese Davis, "Remembering Our Ancestors: Cross-cultural Collaborations and the-Mediation of Aboriginal Culture and History in *Ten Canoes*," *Studies in Australasian Cinema* 1, no. 1 (2007): 5–14. Hereafter cited as *TCS*.

5. *Ten Canoes*, "Press Kit," Palace Films Australia, 2006, last modified July 1, 2010, http://www.tencanoes.com.au/tencanoes/info.htm, hereafter cited as *TCK*.

6. Louise Hamby, "A Question of Time: *Ten Canoes*," *The Australian Journal of Anthropology*, 18:1 (2007): 123–26. Hereafter cited as *AJA*.

7. Lindy Allen. Personal Communication with the author (May 2008).

8. Nicolas Peterson, Introduction to *Donald Thomson in Arnhem Land* (Melbourne: Miegunyah Press, 2006). Hereafter cited as *DTA*.

9. Gaynor Macdonald, "Photos in Wiradjuri Biscuit Tins: Negotiating and Validating Colonial Histories," *Oceania* 73 (2003): 225–42.

10. Philippa Hawker, "Canoe Culture Bridges Gap," *The Age*, June 3, 2006, 18.

11. Lindy Allen, "Tons and Tons of Valuable Material," in *The Makers and Making of Indigenous Australian Museum Collections*, ed. Nicolas Peterson, Lindy Allen and Louise Hamby (Melbourne: Melbourne University Press, 2008). Hereafter cited as *IAM*.

12. Roslyn Poignant and Axel Poignant, *Encounter at Nagalarramba* (Canberra: National Library of Australia, 1996).

Chapter Nine

Mapping Modernity in "Marvellous Melbourne"

Ada Cambridge's A Woman's Friendship

Harriet Edquist

Ada Cambridge (1844–1926) published her novel *A Woman's Friendship* in the Melbourne *Age* in weekly instalments from August to October 1889.[1] Cambridge had established herself as a fiction writer in the 1870s and achieved great acclaim with the publication of *A Marked Man* in 1890. Her genre was romantic fiction and critics generally deem her themes to be narrow; "there is really only one—love (or, in a vulgar sense, romance). The actors are women, and their desire or fate or accident is to marry" (*ACT*, 12). If this is the case, *A Woman's Friendship* is not one of her customary tales.

INTRODUCTION

The central characters in *A Woman's Friendship* are two women in their thirties, Patty Kinnaird, the wife of a sheep farmer and Margaret Clive, the wife of the editor of a city newspaper. Having met on a boat trip from Sydney they become friends and establish a Reform Club, an informal reading group intended to discuss books and debate the latest ideas. The 1888 Centennial Exhibition in Melbourne provides the occasion for their meetings as Patty and her husband Ted come to town on extended visits. In time they invite the seemingly sympathetic Seaton Macdonald, a wealthy landowner and well-born member of Melbourne society, to join them, to prove friendship between the sexes can be Platonic and achieved without sexual jealousy. In vain, of course, as Macdonald, a practiced womanizer, expertly insinuates himself into the affections of each, plays them off against each other and sets out to seduce Patty. The two women part, their friendship ruined until eventually they see Macdonald for what he is and are reconciled, better friends than before.

If we examine the novel along the traditional lines of plot and character, we might agree with those who believe its primary theme is friendship in its many forms. In her introduction to the 1988 edition Elizabeth Morrison argues that the core of the novel is exposure of the sham of their "purely intellectual friendship" with Macdonald (*WF*, xxxiii). Her attitude, and that of Cambridge's biographer Audrey Tate, is strikingly more severe on the women than on the men of the novel.[2] Today, we might be rather inclined to see not the failure of Patty and Margaret in their attempts at self-education and reform, but that of the patriarchal establishment represented by Macdonald, who sets out deliberately to usurp and then destroy the women's aspirations. That they survive his attempt is a triumph for them. Their Reform Club followed the London trend in women's clubs although its name would have brought to mind the Reform Club in Pall Mall, the politically inclined, all-male club founded in 1836 which was housed in an imposing Renaissance revival palazzo designed by eminent British architect Charles Barry. The Melbourne Reform Club, by contrast, was initially restricted in its membership to the two women before it added Macdonald; it was domestic in setting or peripatetic, and interested not so much in the great reform issues of the day that occupied men, but those that occupied women. Its reading list provided the subject matter for spirited debate and included Milton, Mary Wollstonecraft on *Marriage and the Rights of Women*, Matthew Arnold and William Morris, Henry David Thoreau, James Cotter Morison and George Meredith, Cambridge's own favourite novelist. Its three members debate current political questions over dinner in an atmosphere of engaged enthusiasm. If Cambridge's editor David Syme hated the novel, expunged bits of it, and wrote, "This tale is the dullest & stupidest Mrs Cross ever penned & had it not been extensively advertised beforehand I shd [*sic*] have paid the money for it & kept it out of the paper altogether," it was quite possibly because of its advocacy for women's reform rather than fear of libel as Morrison suggests (*WF*, xxxix).

However, if we examine the novel from a different point of view, one provided by the more recent discipline of literary geography, we arrive at somewhat different conclusions. Following the methodology of Franco Moretti's *Atlas of the European Novel*, I "mapped" where action actually takes place in the novel.[3] Temporally, it occupies the six-month period of the Melbourne Centennial International Exhibition, which ran from August 1888 to January 1891. Spatially its main action is distributed across three sites, not just the Exhibition which is the site critics tend to remember and comment on. Of equal importance is the city itself and the domestic interiors of the three main protagonists. In each of these sites the two women negotiate a rapidly changing metropolitan culture and define for themselves what it is to be modern.

THE CITY

"Marvellous Melbourne," the epithet for the city coined by visiting English journalist George Sala in 1885, was an entirely modern phenomenon.[4] By the 1870s Melbourne was, thanks to the gold rush, the Australian center of finance and trade, which in turn attracted British finance which "poured into Victoria in the 1880s attracted by higher rates of interest than could be obtained at home. Both farming and industry boomed . . . The growth of a centralised railways system favoured it economically; the willingness of rich Victorian farmers to spend their money in Melbourne . . . favoured it both economically and socially."[5] Asa Briggs, who included Melbourne in his study of Victorian cities, also notes that Melbourne was proud of its urbanity and that there was a "feeling that there was a distinctive Australian future . . . in the cities [which] were believed to have a superiority of their own." He also notes, that "at their best, they felt that they were making history" (*VC*, 294–95).

Briggs quotes "eminent Victorian" Harriet Martineau who wrote that "whenever she went to a strange city she went at once to the highest point in the neighborhood from which she could see the city as a 'living map' below her." He notes the Victorian enthusiasm for ballooning and saw the aerial view as an analogy for, or a way of synthesizing facts about cities in order to construct "a sense of unity and order" (*VC*, 57–8). Cambridge's characters travel in a hydraulic lift to the highest point of the city—the dome of the Exhibition building—in order to view the "living map" below them. But Cambridge herself synthesizes the facts of the city giving them unity and order in another way at the beginning of the novel. Having introduced her three main characters and the nature of their friendship centred on their Reform Club, she describes the peripatetic way in which they, like Aristotle, hold their discussions:

> They held their first full meeting at Mrs Clive's house [. . . in] East Melbourne . . . the second took place at the Imperial Coffee Palace. . . . Thrice they met in the Public Library; once in the Botanic Gardens; twice in the University museum; four times they went down to Brighton and sat on the beach, and once they took the "Ozone" to Sorrento. But the regular rendezvous was, of course, the great building in Carlton gardens; it was there they had an almost daily symposium, and all that seemed worth remembering of the summer was associated with it (*WF*, 9).

To locate these sites is to construct a synoptic view or map of Melbourne East-West from the Imperial Coffee Palace in Collins Street to East Melbourne by way of the Melbourne Club (where Seaton Macdonald stayed when in town) and North-South from the Botanical Gardens to the University by way of the Public Library. The Exhibition Building in the Carlton Gardens was just off this central urban grid. The reference to Brighton alerts the reader to Melbourne's position on Port Phillip, and to Sorrento its aspirations to an increasingly global resort culture. Thus the locales of the Reform Club meetings map the city's main thoroughfares and institutions of public instruction and amusement; in the compass of one paragraph Cambridge has provided a synthesis of Melbourne as a modern metropolis. We also discover this city is serviced by trains and trams, has well developed communications that allow characters to telegraph each other daily to keep in touch and has at its centre in Collins Street, Clive's newspaper offices, a hub of modern communication. As Graeme Davison points out "few features of the boom metropolis were more conspicuous than the proliferation of exchanges, agencies, trade journals, telegraph and telephone services . . . and other agencies of secondary communication" (*RFM*, 132–33). Although Melbourne was no Paris it had the character of the metropolis which Charles Baudelaire's artist M. G. portrayed so successfully:

> The crowd is his domain, just as the air is the bird's, and water that of the fish. His passion and his profession is to merge with the crowd. For the perfect idler, for the passionate observer it becomes an immense source of enjoyment to establish his dwelling in the throng, in the ebb and flow, the bustle, the fleeting and the infinite. To be away from home and yet to feel at home anywhere; to see the world, to be at the very centre of the world, and yet to be unseen of the world, such are some of the minor pleasures of those independent, intense and impartial spirits, who do not lend themselves easily to linguistic definitions.[6]

In her Melbourne novels Cambridge portrayed the life of the city in its most intense and characteristic moments. In *The Three Miss Kings* (serialized in 1883) one of these settings is the 1880 Melbourne Cup at Flemington where the aptly named Grand Flaneur won the race before a huge crowd of 100,000 people; in *A Woman's Friendship* it is the Centennial Exhibition, that extravagant spectacle of metropolitan culture.

THE EXHIBITION

Situated in the Carlton Gardens on the edge of the city, the Exhibition was housed in Reed and Barnes's colossal, electrically-lit building designed for the 1880 Exhibition, extended and enhanced on the proceeds and expectations of the long boom years. Since the Great Exhibition of 1851 in Hyde Park, London, world exhibitions produced an increasingly globalised modernity fixated on trade, "mass-production, prefabrication, mass communications and urbanisation" (*SAH*, 7). They were the exemplary cultural institutions produced by the nexus between industry and empire that reflected "the driving forces behind Western society up to the Second World War" (*EVE*, 52). With their huge assemblies of visitors and tourists they also reified the spectacle of modernity, and Cambridge depicts her characters as modern flaneurs, spectators of the new world opening up to them, large enough for them to find "solitude in an alien crowd" (*WF*, 52). Susan Martin argues that the Exhibition as a site of imperial culture provides a space where Patty and Margaret are able to "produce a white Australia settler identity through careful sorting—adapting the experience and offerings to their own needs." As opposed to other textual reproductions of the exhibition such as catalogs and newspaper reviews, the novel is "highly selective" in its descriptions.[7] Although Patty is a countrywoman neither she nor her companion pays any heed to the vast displays of primary produce which underwrote the wealth of the colony. Rather, they meet in the places where modern culture can be experienced: the picture courts where they debate the value of popular paintings, and the concert hall where they eventually learn to love the most modern of all composers, Richard Wagner. While the Exhibition is a great democratic meeting place and site of dalliance and seduction, it is also a place of education and Margaret grasps it avidly, seeking to develop her own response to the world around her, refuting popular taste whether it be in painting, music, or women's work, "crochet and the wax flower work and the Berlin wool and crewel work," which to her were "so many gory corpses" (*WF*, 53). While Margaret might appear naive she has, as does her creator and alter ego, Cambridge, some of those traits of Baudelaire's painter of modern life:

> this solitary mortal endowed with an active imagination, always roaming the great desert of men, has a nobler aim than that of the pure idler, a more general aim, other than the fleeting pleasure of circumstance. He is looking for that indefinable something we may be allowed to call "modernity," for want of a better term to express the idea in question (*SWA*).

THE HOME

With their acres of exhibits, the Exhibition brought new ideas about design and style to Melbournians. Cambridge observed: "our famous International Exhibition of 1880 . . . first taught us as a community the rudiments of modern art" and she charts its progress through her novels (*TYA*, 135).

The first domestic interiors, which play host to the Reform Club, are the rooms that Patty and her husband Ted have taken in the Imperial Coffee Palace. The "Imperial" is Cambridge's toponym for the Federal Coffee Palace on the Southwest corner of King Street and Collins Street. This imposing building, an unlikely monument to temperance, was designed by William Pitt and Ellerker and Kilburn in an extravagant eclectic classicism suited to its times. Designed for out-of-town visitors from the country, it was built in anticipation of the Exhibition. Five of its seven stories contained the bedrooms while the ground and first floors housed the public domain—the dining, lounge, sitting, smoking, writing, and billiard rooms. It was full of modern amenities including lifts, gaslights, electric service bells, and it came to epitomize the booming city. The rooms occupied by Patty and Ted were "smart and prim, and not very comfortable" with their "gorgeously upholstered furniture" but Patty liked comfort and "had introduced a couple of softly cushioned homely wicker chairs, which altered its character entirely" and Ted for his part had introduced whisky (*WF*, 27).

Patty's rooms suggest taste comfort and informal domesticity; Margaret's domestic arrangements are the setting for a truly unusual kind of life. A colonial bungalow of the simplest kind built in pre-gold Melbourne situated in East Melbourne, an old suburb close to the city, it is a metaphor for the couple's married life: "Amid streets full of smart white terrace houses, Mrs. Clive's one-storied, shingle-roofed, antiquated detached cottage, in its rose bushy garden, was considered by most people to be conspicuously out of place" (*WF*, 38). Her neighbors would prefer that its owner and the house be removed "from the path of progress." For Margaret Clive however, the choice to live there was a deliberate one, her stand against "a city full of barbarians" in which her house was "an oasis of good taste in a desert of genteel vulgarity." Her taste in this instance mirrors that of Cambridge whose seventh Victorian house at Beechworth (the place where she wrote the novel) was a one-story bungalow in an overgrown garden, "a trifle dilapidated" with all its paint gone and "the soft grey of the dissolving wood-work . . . in perfect harmony with every other detail of the composition" (*TYA*, 190).

In its delight in nature, the description of the exterior of the Clive's house brings to mind the precepts of design reformers, John Ruskin and William Morris:

> . . . the little garden was sweet and gay. A lilac tree just coming into bloom, a clump of red-brown stocks, a bush of boronia, a bed of violets, exhaled the delicious breath of spring; and the flush of living color [*sic*] in red japonica and yellow jonquil and white feathers of spiraea, and in the fringe of wisteria hanging from the low eaves of the verandah, lit up this one only of all the front gardens in that section of the street (*WF*, 38).

The emphasis on scent and color, freshness and informality, the intimate "fringe" of wisteria, charge this passage with Ruskinian naturalness. Ruskin had written "I know what it is to live in a cottage with a deal floor and roof . . . and I know it to be in many respects healthier and happier than living between a turkey carpet and gilded ceiling, beside a steel grate and polished fender."[8]

Interior spaces are among the key sites of action in Cambridge novels and they reflect the emergence of the modern interior in the late nineteenth century. The interior of the Clive house was modern and, although the "aesthetic spirit presided over all the arrangements," it was not a slave to any style (*WF*, 38). It was in fact something new. As Penny Sparke has argued, the Aesthetic Movement marked a "shift from the eclectic, mid-Victorian, morally-oriented, domestic interior to the 'artistic home' [and] was hugely significant in the evolution of the modern interior." It brought with it "a new emphasis upon visuality over materiality and spatiality" and it was in these new interiors that "women learnt how to engage with the rules of taste and fashion," with the emerging notion of a "modern lifestyle" and by these means "form their modern identities" (*MI*, 35–6). By 1889 however, the Aestheticism was on the wane and something new and more radical was slowly emerging, and this is reflected in Margaret's house:

> No swathes of Liberty silk, no festoons of kalizoic muslin adorned that little drawingroom where Margaret received her friend with literally open arms. There was not a china plate anywhere, nor a Japanese fan—none of the familiar gimcracks. She was above the frivolities of fashion in pretty things as in ugly ones; nothing was there for the mere sake of showing itself except the pictures, which were a singularly interesting collection (*WF*, 38).

Indeed the room had an austerity and absence of clutter that marked it out from even the smartest Melbourne town house and being both spare and elegant was reflective of the personality of the owner:

> The dark floor was almost bare, the furniture of a capacious and substantial type not designed for the apartments of ladies, and leaving an unusual amount of space for moving about in, but everywhere color and form and harmoniousness had been studied. The delicate, austere simplicity of the whole was the quintessence of refinement (*WF*, 38).

The picture collection was small but good and the dado comprised shelves filled with books from which the Reform Club took their reading: "Thoreau one day, Cotter Morison the next—always an author of revolutionary ideas" (*WF*, 39). The description of the drawing room betrays knowledge of William Morris's reformist views on art, design and socialism. In one of his most famous lectures, "The Beauty of Life" delivered in 1880, he proclaimed what came to be one of the central tenets of the Arts and Crafts Movement: "Have nothing in your houses that you do not know to be useful, or believe to be beautiful" (*PVL*, 561). Morris illustrated this dictum with a description of a sitting room furnished for a "healthy person":

> First a book-case with a great many books in it: next a table that will keep steady when you write or work at it: then several chairs that you can move . . . you will want pictures or engravings, such as you can afford, only not stop-gaps, but real works of art on the wall . . . Then there will be a fireplace of course (*PVL*, 561).

Although Morris nominated one or two other pieces of acceptable furniture such as a window seat and cupboard, his description of simplicity is that of Margaret's house and in the Clive's domestic setting we see the ideals of the Arts and Crafts movement only then beginning to emerge in Melbourne. It became a force in the 1890s and ushered in the new interior around the turn of the century.[9]

In the individuality and rigour of Margaret's restrained taste Cambridge makes a subtle differentiation from Seaton McDonald whose refinement of perfect taste is overt and somewhat overpowering. Macdonald's house Yarrock at Mount Macedon is the setting for the novel's climax as it is here that the three members of the Reform Club retreat for a weekend away from Melbourne. Occupying the central portion of the novel, this episode is set in the most elaborate of the novel's interiors. Mt. Macedon, the "Simla of the South," was in the late nineteenth century the place to which fashionable Melbourne retreated from the summer heat and Cambridge knew it well having stayed at Charles Ryan's house, one of the models of Yarrock (*TYA*, 169). Yarrock's organic growth and picturesque assemblage of disparate parts over time conformed to the archetype of Cambridge's perfect dwelling. The interior was a triumph of the Aestheticism, perfect in its way and it captured Margaret's imagination, colored as this was by dreams of South Kensington. However, like its owner, Yarrock's fastidious perfection masked an unnaturalness and, like a stage set, its use was primarily for Macdonald's seductions.

In these three interiors, the relaxed comfort of the Kinnaird's rooms at the Coffee Palace, the austere intellectualism of East Melbourne and the over-refined aestheticism of Yarrock, Cambridge presents us with places that reflect the personalities and desires of their makers and shape the activity that takes place within their walls. Action is thus predicated upon place. At the same time Cambridge uses these interiors to argue against the grandiose domestic consumerism of the boom years as she attempts to construct a discourse of modernity where taste reflects not status and money but knowledge of design. Margaret's achievement is to produce a synthesis of advanced contemporary thinking about women's reform, education, and design, and bring it to life in a plausible way. She is intense, and in her surroundings and daily routines, she attempts to live a life of intelligence and truthfulness even if Macdonald temporarily unhinges this.

THE FASHIONABLE BODY

The fourth site on which Cambridge maps the trials of modernity is that of the female body itself. While Patty is conventional in her taste in clothes, Margaret is not; and through her clothes, she attempts to fashion a persona in keeping with her developing ideas on modernity. Margaret's body thus is the visible sign of her beliefs. When we first meet her, she is waiting for Patty at the Exhibition, a conspicuous but not dowdy figure in the crowd:

> A tall woman, with a thin, rapt face, and an air of natural dignity and distinction that enabled her to wear an unfashionable dress without degrading herself in the public eye. She had on a dark woollen gown that had a loose body with a mere apology for a waist and a skirt without a scrap of bustle; and her bonnet, instead of soaring half a yard into the air, touched its highest point about 3 inches above her head (*WF*, 11).

Her appearance may have been modeled on that of Harriet Dugdale, a pioneer of the Women's Movement in Victoria, President of the first Victorian Women's Suffrage Society in 1884 and advocate of evolutionary progress, rationalism, and dress reform; she modelled her clothes on those of American dress reformer Amelia Bloomer.[10] Dugdale was also a vegetarian and although Cambridge does not explicitly state it, Margaret was as well. She disliked meat on principle and "shudderingly termed [it] 'flesh,' as if to like it were a sort of cannibalism" (*WF*, 47). Cambridge herself was not a vegetarian, but dress reform did interest her and when an invalid she seized the opportunity to wear "the most delightful costume that I ever wore in my life"—"a long, light, loose paletot of China silk . . . buttoned at the throat and all down the front to the hem, which cleared the ground by about three inches. It had roomy pockets outside; the sleeves were roomy also; there was no need to wear a dress under it, nor anything whatever round the waist . . . Night after night in this delicious liberty, we roamed the city everywhere" (*TYA*, 172).

Macdonald, seeking to out-maneuver Margaret and her developing ideas on women's reform has already published a paper on "The Liberty of Woman" in a Melbourne newspaper and intended to follow this up with one on "The Relation of Women's Clothes to their Moral and Intellectual Development" (*WF*, 62). The title of this paper probably refers to Mary Haweis' "Moralities of Dress" in *The Art of Beauty* (1878).[11] The English Haweis, like Cambridge married to a cleric, was an indefatigable writer on art, taste, dress and interior decoration. While typically mid-Victorian in many of her attitudes she was also an advocate for dress and design reform and brought ideas from Morris into the debates around contemporary fashion, particularly those to do with simplicity, truth to form and healthfulness. When we first meet Margaret in the sitting room of her house in East Melbourne she is "in a yellowish Greek-draped dress that beautifully adapted itself to her dark head and gentle majesty of movement . . . " (*WF*, 38). Haweis advocated the Greek style for its healthy, uncorsetted naturalness, which allowed great freedom of movement. She also drew attention to the way in which the body should complement the room in which it is placed (*AB*, 224–33). Cambridge follows suit in this passage:

> She leaned back in her big chair. The faint grey blue of the chintz that covered it made a good background for the rich darkness of her hair and her liquid eyes. With her face to the firelight she looked beautiful in her tawny draperies, lying at rest in that flexible, curving pose, with the red glow flushing her all over (*WF*, 38).

Haweis devotes several chapters of *The Art of Beauty* to the relationship between a person and their surroundings through an analysis of color harmonies and it is clear that Cambridge arranges her figures in their interiors with these rules in mind; Margaret looks beautiful in this interior, but wan and old outside; Yarrock's interiors are so designed to make a woman of her coloring youthful and beautiful, and so on.

The prelude to the trip to Mt. Macedon takes place in Margaret's bedroom which is "a severely simple, carpetless, curtainless, but extremely dainty apartment." Here, she and Patty who is staying with her (Ted having returned to the country) select their wardrobe. Margaret attempts to persuade Patty to disavow her bustle and wear a garment constructed from a "finely textured oriental window curtain" and draped about her body in the Greek style. Patty, who is a realist, sees the beauty of the fabric, its lovely fold "sweeping in free lines to the floor" but also how "classic robes were more becoming to a tall and willowy figure like Margaret's than to one so short as hers" (*WF*, 62). While she reluctantly assents to wear the robe, she secretly packs her conservative, bustled black evening dress, which she knows sets off her figure and will be attractive to Macdonald. When the moment of trial comes at Yarrock, and Margaret is dressed so that "no unprejudiced and cultured eye could have seen her when her toilet was completed without acknowledging that the Greek dress was fully as beautiful as she believed it," Patty had dressed for Macdonald:

> She rustled into the passage, shaking out her train behind her, lifting a heavy curtain, found herself in the hall. There stood Mr. Macdonald on the hearthrug,

> in evening dress, with his back to the fire, and, whatever might have been his passion for classic raiment, and his abhorrence of the current mode, he certainly regarded her with an eye of admiration as she stepped towards him. She knew the look too well to make any mistake about it, and congratulated herself with fervour on having chosen to be modern, after all (*WF*, 74).

For Patty, modernity is equated to fashion and convention; for Margaret, modernity in dress is equated to the freedom of women to express themselves as they wish and to liberate their body from the corset. For Mrs. Haweis, "Dress bears the same relation to the body as speech does to the brain; and therefore dress may be called the speech of the body." She claims two general rules to be observed in dress. Firstly, "it shall not contradict or falsify the natural lines of the body" and secondly, "the attire shall express to a reasonable extent the character of the wearer" (*AB*, 23). While Patty fails the first test, she passes the second. Margaret passes both.

CONCLUSION

To focus attention on the settings for action in *A Woman's Friendship* allows us to see the narrative in a new light. While the plot is driven by the doings of the Reform Club, the settings tell a different story. While a reading of the primary narrative might lead us to focus on the urban environment, an examination of the spatial narrative shows that the domestic interior is of more significance. In the primary, plot-driven narrative the women are seduced and betrayed by Macdonald and all he represents; in the secondary spatial narrative, Margaret is triumphant. Her resistance to the status quo, her enthusiasm for the stimulation of metropolitan pleasure but insistence on clearly marking out within it her own territory both physical and intellectual, brings to mind some of the tensions recognised in metropolitan life by Georg Simmel in *The Metropolis and Mental Life*:

> the metropolis is the genuine arena of this culture which outgrows all personal life. Here in buildings and educational institutions, in the wonders and comforts of space-conquering technology, in the formations of community life, and in the visible institutions of the state, is offered such an overwhelming fullness of crystallized and impersonalized spirit that the personality, so to speak, cannot maintain itself under its impact. On the one hand, life is made infinitely easy for the personality in that stimulations, interests, uses of time, and consciousness are offered to it from all sides. They carry the person as if in a stream, and one needs hardly to swim for oneself. On the other hand, however, life is composed more and more of these impersonal contents and offerings which tend to displace the genuine personal colorations and incomparabilities. This results in the individual's summoning the utmost in uniqueness and particularization, in order to preserve his most personal core. He has to exaggerate this personal element in order to remain audible even to himself.[12]

In her reformist zeal perhaps Margaret understands the need to "exaggerate this personal element in order to remain audible even to herself." In her attitude to fashion as in all else, Margaret represents Simmel's duality between social conformity and individual expression. Behind the veil of irony that Cambridge sets between herself and her heroine lies a very modern proposition.

NOTES

1. Ada Cambridge, *A Woman's Friendship*, ed. Elizabeth Morrison, (Kensington: New South Wales University Press 1988). Hereafter cited as *WF*.

2. Audrey Tate, *Ada Cambridge. Her Life and Work 1844–1926* (Melbourne: Melbourne University Press, 1991).

3. Franco Moretti, *Atlas of the European Novel* (Verso, 1998). See also Harriet Edquist, "Precise and Imprecise Geographies in Christina Stead's *Seven Poor Men of Sydney*," *The Cartographic Journal* 46, no. 4 (November 2009): 343–49.

4. "George Augustus Henry Sala," Wikipedia, accessed November 15, 2009 http://en.wikipedia.org/wiki/George_Augustus_Henry_Sala.

5. Asa Briggs, *Victorian Cities* (Harmondsworth: Penguin, 1985), 287. Hereafter cited as *VC*.

6. P. E. Charvet, trans., *Charles Baudelaire: Selected Writings on Art and Literature* (New York: Viking, 1972), 395–422, in "The Painter of Modern Life. Charles Baudelaire (1863)," Modernism och postmodernism, last modified December 22, 2010, accessed November 15, 2009, http://www.idehist.uu.se/distans/ilmh/pm/baudelaire-painter.htm. Hereafter cited as *SWA*.

7. Susan Martin, "'Surmounted by Stuffed Sheep' Exhibitions and Empire in Nineteenth-century Australian Women's Fiction," in *Seize the Day. Exhibition, Australia and the World*, ed. Kate Darian-Smith et al. (Clayton: Monash University ePress 1 no. 1 September 2008), 13.7.

8. Quoted in Penny Sparke, *The Modern Interior* (London: Reaktion Books, 2008), 22. Hereafter cited as *MI*.

9. Harriet Edquist, *Pioneers of Modernism. The Arts and Crafts Movement in Australia* (Melbourne: Miegunyah Press, 2008).

10. Janice N. Brownfoot, "Dugdale, Henrietta Augusta (Harriet) (1826?–1918)," accessed November 15, 2009, http://adbonline.anu.edu.au/biogs/A040111b.htm.

11. Mary Eliza Haweis, *The Art of Beauty* (Boston: Adamant Media, 1878/2005). Hereafter cited as *AB*.

12. Georg Simmel, "The Metropolis and Mental Life 1903," adapted by D. Weinstein, *The Sociology of Georg Simmel*, trans. Kurt Wolff (New York: Free Press, 1950),409–24, accessed November 15, 2009, http://www.altruists.org/f792.

III

Mapping a New Biological Paradigm for Design

Introduction

Hélène Frichot

Bioarchitectures are becoming globally ubiquitous in architecture and design schools as well as in digitally or computationally adept architecture and design studios or "laboratories." Now that digital representation is commonplace, and computational processes such as parametrics, the use of evolutionary algorithms and associated scripting increasingly enter the design process, a return to organic form and the underlying code that generates life has been ventured. The chapters in part III of *De-signing Design* map the emergence of what can be identified as a new biotechnological paradigm for architecture and design. Techniques associated with contemporary experimental architectures undertaken in what could be called a post-digital milieu are drawing increasingly on the now well-established cross-fertilization of ideas between computer science and biological science. If we were to render today a list of symptoms that pertain to the current engagement of the discipline of architecture with the above conjunction, conceived as a *biotechnogenesis* of the ever-evolving human condition, we would find that the large question of life, and a vested relationship between architecture and life, recurs across contemporary architectural discourse and production at a global scale. The writers in part III address what is at stake in the complicated encounters between architecture, design, and life. While relying on a range of disciplines and approaches, they venture into a problematic field that resides between digital design and the life sciences, examining how techniques and transfers of knowledge have impacted on modes of design discourse, expression, materiality, and production.

Although histories of architecture and design reveal marvelous displays of the various means by which natural forms have been emulated through the force of human labor, what is remarkable about contemporary developments is the focus not only on biological process over the conceptual illusion of reified natural form, but also on the suggestion that architecture will become living, ever-transforming, or morphogenetic organism in profound, symbiotic relationship with environmental context or ecological niche. Biomimetic metaphor has given way to biological metamorphosis. It is necessary to turn our attention to the microscopic, molecular, and even atomic scale of a world to comprehend possibilities of continuity between what, at the mere human scale of things, seems to constellate in patterns of organic and inorganic array.

Recent collaborations such as the Emergence and Design Group (Michael Hensel, Michael Weinstock, and Achim Menges), have begun to explore how architecture, one day soon, will literally respond to life criteria. According to Michael Hensel and Achim Menges, we have much to learn from the chemical reactions that occur across the flexible membranes of cell walls, as well as their associated material infrastructures.

In perusing the pages of *Architectural Design (AD): Emergence: Morphogenetic Design Strategies* (2004) and *AD: Techniques and Technologies* (2006), images of architectural form and life forms in magnified detail are placed alongside each other in an undifferentiated manner. Yet the close vision (literally life magnified) of the Emergence and Design Group threatens a collapse with respect to our ability to create what might turn out to be useful differentiations or alternative and critical approaches to emerging bio-architectures.

Hélène Frichot's opening chapter, "Mapping an Ethico-Aesthetics for Wet Biotechnological Architectures," conceptually maps and frames many of the issues raised in the chapters gathered in part III of *De-signing Design*. This chapter surveys a series of examples of biotechnogical design both local (to Melbourne) and global, focusing upon the increasing deployment of theories of emergence. The increasing indistinction in contemporary digital design discourse between defining what is architectural form and what is living form is also addressed. The stability of architectural form has given way to the vicissitudes of open-ended design process, which is often aligned with life processes. The chapter assesses speculations on a future for architecture conceived as a form of artificial life and will touch briefly on the theme of the monstrous as a limit condition. The questions include: What is at stake in the artificial animation of architecture? What practical ethics might be engaged such that these experiments augment rather diminish the continuance of life forms that are ever in formation, whether individual, social, human, architectural, or other?

Karen Burns's chapter, "Digital Organic Design: Architecture, the New Biology and the Knowledge Economy" continues the discussion with a specific focus on the work of architect and theorist, Greg Lynn. Animate form, blobitecture (organic form), and the fold as conceptually delivered to us through the writings of French philosopher, Gilles Deleuze, are the preoccupations that frame much of Lynn's career. Lynn's introduction to the thought of Deleuze in the popular architecture journal, *AD* in the 1993 edition *Folding in Architecture*, opened a way into arguing for pliable form that countered the aggressive shards and angles that had come to characterise so-called deconstructive architecture. *Folding in Architecture* heralded a shift in the signs that constellate to form the architectural theory and practice firmament. Burns focuses in particular on Lynn's Embryological House, a prototype of which was exhibited at the Centre Pompidou in Paris between 2003–2004 as part of the *Architectures Non Standard* exhibition curated by Frédéric Migayrou with the assistance of Zeynep Mennan. Burns carefully critiques the evolutionary theory that Lynn invests in his architectural design process and what she calls Lynn's "bio-discourse." The essay outlines sites of tension and difference that arise between evolutionary theory and architectural theory, and shows how by recourse to scientific metaphors architecture seeks to secure claims for authority within its own domain and with respect to its own creative enterprise.

Stephen Loo, in "De-signing as Bio-Technological Endosymbiosis" introduces us to the close vision of biological life genetically unfolding in order to argue that design can be aligned with biology not so much formally, but according to a systemic framework. To forward his argument Loo draws predominantly on the influential work of Deleuze and Guattari, many of whose concepts have been imported into architectural design discourse and process, from the 1980s onwards, in the early stages of what we now identify as a new biotechnological paradigm in architecture. The essay discusses a relationship between biology and technology, and suggests that these are reciprocal co-producers in their respective evolutions. After thinkers such as Georges Canguilhem and Gilbert Simondon, Loo asks that we reconsider how we tend to frame both machines and biological life forms in terms of parts in relations to wholes.

Living beings cannot be reduced to technological machines, and furthermore, technological machines are perhaps, after all, systemically closer to biological beings in terms of their development through historical time. Technological systems, like living systems, can be explained in terms of growth, differentiation, redundancies, and so forth. What information is to technology, genetics is to living systems, and both continue to vary in response to the environmental contingencies and vicissitudes presented to them.

Lisa Dethridge explores the virtual life world that is "Second Life," and ventures into a mode of existence that favours simulation and the freedoms of disembodiment. Once we begin to imagine posthuman futures that rise up as the result of new combinations between technologies and biologies, then one possible direction into which such futures lead us is these virtual or electronic realms wherein the meat of the body is merely that which supports the active life lived on line.

The theme of the monstrous, ventured in Frichot's discussion, is addressed in more depth in Chris Smith's contribution, "Hopeful: Biology, Architectural Design and Philosophy." Smith shows us how life is inherently wayward, and that design can sometimes contribute to monstrous outcomes. There is not only horror but also wonder to be discovered in signs of the monstrous. This chapter also draws on the philosophy of Deleuze and Guattari to elaborate on how identity is never decided once and for all, but open to the necessity of unexpected encounters. As Smith explains this requires a mode of thinking that emphasises the force of existing over stable form. The figure of the monster is exemplary as it is a body that is composed of material and semiotic connections; it is an assemblage of protean parts and mobile relations. The monster, as a figure, de-signs design as it disrupts expectations and challenges the idea of a norm with the expression of the exceptional. It also traverses disciplinary boundaries with ease allowing fiction, philosophy, biology, and architectural design to be brought together in a provisional way. In this chapter, Smith specifically draws his monster from the field of architecture with all its "excrescences and orifices."

Neil Leach closes part III, and ends the whole collection, with a discussion of what he calls "New Materialism." The increasing sophistication in digital technologies and the designer's grasp and understanding of computation has not resulted simply in a headlong dive into virtual worlds, such as those described by Lisa Dethridge in chapter 13. An aptitude in the management of new digital design technologies has also folded back into very real material implications.

It was Manual DeLanda who first coined the notion of a new materialism by drawing predominantly on the philosophy of Deleuze and Guattari. Much of DeLanda's work has sought to explain aspects of Deleuze and Guattari's conceptual project in relation to the scientific frameworks that have inspired their philosophy. New materialism forwards a non-hylemorphic understanding of matter, that is, matter is determined no longer by a form or idea that imposes fixed constraints upon it, making it inert. Instead, matter is conceived as participating in an open and dynamic system that is self-organized from the bottom-up, and as such exhibits emergent properties. The implication of this is a de-emphasis on the visual and formal concerns of architecture, and an emphasis on how architecture performs in response to its given environment. Leach identifies this shift as both a new paradigm and a performative turn in architectural design culture, as well as a shift away from theory based in philosophical influence, and toward a positivistic theory based in science, technology, and material behavior.

The question of what is at stake in the complicated encounter between architecture, design and life is what unites the essays collected here. It is necessary to admit that there is nothing new in asking such a question, except that each time we address the question of a life it will have necessarily shifted; new developments in technology and the emergence, as well as the disappearance of ever-new life worlds and life forms demands the reframing of pertinent, context relevant questions. The chapters in the final part of *Designing Design* aim to tackle these ethical and aesthetic questions.

Chapter Ten

Mapping an Ethico-Aesthetics for Wet Biotechnological Architectures

Hélène Frichot

Architects engaged in experimental digital design practice are returning increasingly to the study of life forms for inspiration. Although the so-called natural world has always provided formal tropes to the architect, the underlying processes of biological life, abstracted through the use of algorithms, now drives much design research in the domain of digital or rather, computational architecture. In their biomimetic investigations creative architectural practitioners are advised to equip themselves with a working knowledge of calculus, not to mention evolutionary science, and to remake themselves as technicians—and/or scientists—of an electronic realm of discrete bits, ready to take on genetic algorithmic adventures. A combinatory of computer science and biology has given rise to a term that has by now become a conceptual refrain, "emergence." Defined simply, emergence refers to the way in which basic units, determined by simple rules or codes, assemble in the neighbourhood of each other and begin to behave in such a way as to create complex wholes—for example, swarms of insects, flocks of birds, ant nests and even cities. When applied to the generation of architecture, the surging forth of self-organized life in the emergence of complex and novel systems promises the wonder of built forms that become living organisms, or a wet architecture.

With this chapter I would like to ask: What is at stake in this speculative and artificial animation of architecture? Furthermore, what practical ethics might be engaged such that future experiments augment rather diminish the continuance of life forms that are ever in formation, whether individual, social, human, or other? To venture a tentative direction with respect to the question of a new ethics, or ethico-aesthetics for digital architecture, I will make use of the work of French thinkers, Gilles Deleuze and Félix Guattari. Their ethics of immanence, often overlooked by architects in favor of their implied aesthetics, offers the promise of new social formations rather than the tired tirade of moral rules to be followed. The fruitful conjunction of ethics and aesthetics also suggests that ethics can be actively pursued in creative ways and that an aesthetics can be formulated that is decidedly ethical.

In the field of computational architecture the prevalent ideology privileges quasi-scientific discourse in the articulation of design processes that remain open-ended or process driven. The risk of this discourse is that it reduces spatial problems to a codified regime that reifies rather than augments life—a life, any life whatever. Perhaps what is at stake, and the discursive drift that has returned the question of life to the agenda of architectural discourse, is a somewhat misled desire to return to a real imagined as a paradisiacal nature of endless potential. Here, I will survey a series of contemporary digital design projects drawing together global and local influences (local in this case being Melbourne, Australia) that invest in the combinatory of computer sciences and life sciences. The argument suggests that, obscured in the midst of the novel algorithmic adventures undertaken by computational architects in search of digital artificial life, is the legacy of the work of French philosopher, Deleuze and his occasional collaborator, the psychoanalyst, Guattari. It is predominantly from Guattari that I borrow the ethico-aesthetics of the present chapter's title as a means of ethically, aesthetically, and politically engaging in the frequently non-critical discourse and practice of computational architectures.

Firstly a detour through the work of Hannah Arendt, who is interesting as she stands as a philosopher poised at the brink of much of the work that emerged between computer science and biological science from the 1950s onwards. In her seminal book, *The Human Condition*, Hannah Arendt draws a distinction between *animal laborans* and *Homo Faber*. Where *Homo Faber* is the fabricator of the human world, working with her hands such that the earth is transformed into world, *animal laborans* labors away incessantly in order to sustain the very possibility of the continuance of his life. *Homo Faber* transforms mute material into worldly artifact with her hands, and through this process controls the world in the act of wilful creation. Where *Homo Faber* wants to make life more useful and beautiful, *animal laborans* strives to make life easier and longer (*HC*, 208). Irrespective of these differences, Arendt says:

> not only the utilitarian philosophy of *homo faber* but also the men of action and the lovers of the results of sciences have never tired of pointing out how entirely "useless" thought is—as useless, indeed, as the works of art it inspires [yet] the activity of thinking is as relentless and repetitive as life itself, and the

> question whether thought itself has any meaning at all constitutes the same unanswerable riddle as the question of the meaning of life (*HC*, 170–71).

Arendt also describes an important operational concept that has become even more pertinent of late with respect to architectural design production: process. Arendt describes the interminable, even unstoppable force of process. The concept of process is discussed both in relation to natural processes, and in relation to the fabrication processes by which *Homo Faber* gets things done: "processes, therefore, and not ideas, the models and shapes of the things to be, become the guide for the making and fabricating activities of *homo faber* in the modern age" (*HC*, 300). Following Arendt's account, it is with modernity in particular that process comes to be extracted from nature and turned to technological ends. Where process is identical to the unfolding of the forces of nature, it becomes distinct in relation to the "products of human hands" (*HC*, 150).

With the advent of computation as a crucial part of the representation of architectural projects, and more radically, the ever-evolving action concomitant with design, process can be considered a key term. Despite the transfer of process from natural to techno-scientific ends, the vital life inherent in the notion of process in fact thwarts the best efforts of *Homo Faber*, who, according to Arendt's account, is concerned with ends over means: ends that can be identified in the completed artefact, the monument, the architectural object. An emphasis on process erases the end product as such and sets it into the indefinite circulation of production, consumption, production. This is a preoccupation that has increasingly come to the fore in the practice of computational architecture.

As we will see, digital architecture seeks to de-emphasize the end product in favour of interminable process, thereby exposing a desire to place its activities back on the side of natural, open-ended process. Arendt, who argues that *Homo Faber* has been defeated by *animal laborans*, asks, "why within the diversity of the human condition with its various human capacities it was precisely life that overruled all other considerations" (*HC*, 313). Arendt speculates that exactly through his fascination in capturing, measuring, scientifically accounting for life, man by increments returns to the animal he once was; an animal supposedly emptied out of thought (*HC*, 322). The greatest risk for Arendt is that thought becomes lost to us. At the conclusion of her book she attempts to slow all the fervent activities of the *vita activa*, in order to rest in the shelter of a little contemplation, which she provocatively suggests requires conditions of political freedom in order to flower (*HC*, 324). Arendt is provocative in the 1950s, as she already draws attention to the transformation of life and world through technology, perhaps an age old concern, that nevertheless shifts gear with the conjunction of computation and biological systems (*HC*, 151). While politically astute,

Arendt remains conservative about possibilities for the body, specifically the post-human body, a concept arriving on the scene after her work is done. For Arendt, the body is stable, fixed, and inalienable, its interactions in space and with technology are relatively banal. Arendt asks less what a body can do, and pursues instead the infinite flights of thought, or the contemplative life.

Thought as an active engagement with an immanent milieu is quite a different matter for Deleuze and Guattari. It is a creative activity and not a will to truth; it is dangerous, fraught, "begins to exhibit snarls, squeals, stammers." If thought searches, "it is less in a manner of someone who has a method than that of a dog that seems to be making uncoordinated leaps" (*WP*, 55). The importance of the interplay between the fits and starts of life, and the uncoordinated leaps of thought that Deleuze and Guattari celebrate through the laying out of what they call a plane of immanence and the concomitant construction of concepts will become clearer as I proceed.

To claim a point of view from the present I will now turn to the work of Catherine Ingraham, a well-respected architectural theorist based in New York. In her book, *Architecture, Animal, Human: The Asymmetrical Condition*, Ingraham appears to implicitly confirm Arendt's suggestion that we are overcome with an interest in life, by announcing that "the subject of life always raises the stakes" (*AAH*, 1). When it comes to architecture, Ingraham argues, biological and psychological life are the precondition for the existence of architecture and "must always be indifferent to the life within it:" hence the "asymmetrical condition" of Ingraham's title. Nevertheless, what Ingraham calls "the competition between the wet and the dry and the question of the technological animation of the computer, are already at work in architecture in subtle and overt ways" (*AAH*, 328). I draw the "wetness" of my title from Ingraham's description of this competition, or asymmetry between the wetness of biology and the dryness of computer hardware and software. Ingraham, who is interested in the impact of new computational processes on architecture, maintains an asymmetry between human, animal and other life, on the one hand, and material constraints or framed enclosures of architecture, on the other. Advancing computer technologies, themselves increasingly life-like in their operational capacities, are provoking experimental, avant-garde digital architects to pull down the artificial wall between architectural form and human form and to imagine a continuum that unfolds in both directions. The human and non-human, or animate and inanimate, materials are conceived as interpenetrating one another, for instance, like the hard and soft parts of an oyster and its shell, microscopically generated from the hard and soft layers of the shell that the live oyster secretes.[1]

Apprehending a continuous variation between life and architecture, a symbiotic relationship evolves to suggest all sorts of future possibilities for our understanding of what constitutes human identity, whether it can still be distinguished from that which houses it, and whether its enclosure is, in turn, something other than the life it shelters in its midst. Ingraham confirms that digitally astute practitioners are increasingly fascinated by the potential to bring life movement, vegetative and animal forms in formation into the arena of architectural form making (*AAH*, 328). Ingraham has in mind such architect-theorists as Greg Lynn, who understand that forces are less to do with Newtonian science than a way of understanding the animated field upon which the designer operates, a "new arena for intricate responses in architecture to everything around, in, and of it" (*AAH*, 319). Architectural form, emerging out of a field of forces has thus shifted in kind in that it is no longer stable, but apt to unpredictable transformations that respond both to internal (genetic) rules and external, environmental factors.

Would Ingraham's asymmetrical condition not be better conceived as a continuum of human, animal, other life, organic and inorganic parts, a continuous variation that resembles an undulating field upon which everything gets played out? A field, what's more, that owns a reverse and right side, conjoining both thought and material conjunctions, the wet and the dry of all variety of bodies? On this field, which can be found through various permutations across Deleuze and Guattari's *oeuvre*, thought proceeds in an uncoordinated manner, responding to chance-like shocks and encounters that are, paradoxically, the necessary relations of our composite existences. Arendt's notion of the contemplative life is radicalised by Deleuze and Guattari such that thought exposes us to grave dangers, and far off lands, but importantly, thought is always co-present and co-productive with material admixtures, the affects and percepts of different kinds of bodies, including architectural bodies. Ingraham, for instance, stresses that along with the question of changes wrought on architectural form is the difficulty of accounting for the shifting shape of human life itself. What sense of aliveness is shared between these domains, which we have formerly imagined to be sheltered one inside the other, human form curled in architectural enclosure, soft, wet part inside hard dry shell?

Conventionally, an architectural design process results in an architectural form, built or unbuilt. With digital architecture, as I have argued above, the outcome is de-emphasized in place of the process. Modes of representation shift such that orthogonal drawings no longer account for what is architecturally formulated. Instead, diagrams, animations, fly-throughs better explain the architectural proposition. The question remains, if a pressing contemporary problem is identified, how does the architect seek to resolve this problem? How does a system of continuous variation meet an end, when the end continues to transform? By mimicking life processes, is it that architecture comes to be birthed, learns through life experiences, forgets again, grows old, and slowly dies?

In order to consider some of the questions and problems broached above I will present a series of architectural examples, all of which have a vested interest in the vital signs that might be detected in a computational architecture that increasingly invests in what might be identified as a new biotechnological paradigm.

EXAMPLE ONE: GREG LYNN'S EMBRYOLOGICAL HOUSE

Greg Lynn is a key figure in what could be given as a paradigm shift that takes us from the rectilinear form of the so-called modernist box sitting distinct from a *tabula rasa* field, through the deconstructed, shard-like forms of the nineteen eighties, to the smooth, continuous, and variable form of the so-called "blob" that supposedly emerges out of a given field, or in response to environmental conditions (natural and artificial). In 1999, Lynn designed his Embryological House, which is less a house than a system articulating strategies that respond to issues of customisation, variation, flexible manufacture and assembly, and site specificity. Lynn explains, "there is no ideal or original Embryological House, as every instance is perfect in its mutations" (*CBF*, 92). The Embryological House is not one singular and fixed form, but an open system that allows for an unending series of formal permutations. Of his serial experiments (he formally tested six instances of the house) he says, "I love them all equally as if they were my children. The design problem was not about the house, but the series, the entire infinitesimally extensive and intensive group" (*CBF*, 92). Lynn's anthropomorphic attitude troubles his uptake of the embryological process; he personalizes the process rather than freeing potential forces. There is also the issue of the transgendering that takes place here, in that Lynn acts as mama and papa, superseding the necessity of the maternal womb for the creation of his "children."[2] By basing its inception in the morphogenesis of individual human life, does this architecture assume the same body, and the same regimes of subjectivity that we are familiar with, or does it open up new universes of value, and generate transformative possibilities and modes of expression? It appears to promise the latter, while remaining trapped in the former.

Lynn is also well known for the *Folding in Architecture* edition of *AD*, where he popularised the work of French Philosopher, Gilles Deleuze, translated, sometimes too literally, into the variegated folds of architectural form. There we find that it is less the modernist box, *per se* that Lynn's new architecture promises, than an escape from the tortured forms of deconstructive architecture and the tenuous link made between this architecture and the work of another French philosopher, Jacques Derrida. Lynn explicitly counters the heterogeneous, fragmented and conflicting formal systems that arrive out of deconstructivism, with the supple, pliant, softened and folded curves that are inspired by the work of Deleuze, specifically through his book, *The Fold: Leibniz and the Baroque*.[3] The emphasis, either way, lies in a formal distinction between the rectilinear (however fragmented or broken) and the curvilinear. Lynn announces a further conceptual shift as his work develops and he comes to recognize his growing dependence on computer software and the underlying calculus that supports computational processes. What is most interesting in his progression from an interest in animation, the fold, and the philosophy of Deleuze, to the application of the geometric engines of calculus-based design software, is the underlying fascination in something that approximates vital forces wedded with architecture. As Ingeborg M. Rocker in conversation with Lynn suggests, a further implication of this shift requires that an investment in the history and theory of architecture be placed to the side in favour of "the technological regimes of computational design devices" (*CBF*, 89). This marks a disturbing trend in architectural discourse whereby theory has come to be discarded, making architectural production a techno-scientific activity: architecture risks losing the possibility of framing intellectual projects with any kind of ethical depth.

EXAMPLE TWO: EMERGENCE AND DESIGN GROUP

In 2004 *AD* published *Emergence: Morphogenetic Design Strategies*, edited by the Emergence and Design Group, Michael Hensel, Achim Menges, and Michael Weinstock. In late 2006 there appeared a sequel of sorts, *Techniques and Technologies in Morphogenetic Design*, and in 2008 a further addition to the series, *Versatility and Vicissitude: Performance in Morpho-Ecological Design* (*VV*).[4] Parenthetically, a brief examination of *AD* illustrates the rapidity with which architecture and its associated concepts fall in and out of fashion. For instance, with decreasing frequency the proper noun, Gilles Deleuze can be found across its pages. As Reinhold Martin has explained, it is now with some embarrassment that we venture to name this legendary philosopher (*HDM*, 2). The legacy of Deleuze's thought still, implicitly, inspires a great deal of the content, but what we now read is a language that sounds decidedly quasi-scientific or else directed at the computer scientists among us.

In *Techniques and Technologies*, the Emergence and Design Group describe our human biosphere and venture an enquiry into the consequences of understanding architecture as living entity, as well as the "potential benefits of applying life criteria to architecture" (*TTM*, 6, 17). Life in this context is quantifiable, controlled, mapped, recreated, grown, manipulated, subject to biopolitical measures, and "the entire energy dependent process called 'life' is enabled through photosynthesis" (*TTM*, 23). It is a matter of biomimetically learning lessons from nature, and of assimilating life criteria toward architectural processual ends. The synthetic life of architecture ought to specifically attend to the criteria of containment, metabolism, homeostasis, heredity and evolution. In the pedagogical context of the design studio, it is not uncommon today to hear a student or even certain architects describe how they have "evolved" a form, or else how they have "grown" a form, as though it were a matter of tending one's garden. What's more, there is software available to facilitate such tasks, for example, Surface Evolver, which enables the interactive modelling of liquid surfaces. It is a strange experience to peruse the pages of these three editions of *AD*, because architectural form and natural form, in intricate microscopic detail, stand side by side as siblings. To these collaborations can be added a further publication, *Morpho-Ecologies*, a design manual or user's guide of sorts that announces a new biological paradigm for architectural practitioners.[5] What is remarkable about this addition to the ongoing collaborative project of the Emergence and Design Group is not just the investment being made in biology, but the return of digital techniques to material tests. An interface is elaborated between the electronic and often abstract and scale-less realm of computer aided design and the material manifestation of the processes explored. More recently a further edition of *AD* has been published, this time edited by Achim Menges, entitled *Material Computation*, which suggests that a novel convergence of computation and materialisation is on the brink of emerging. What the biotechnological paradigm apparently allows the designer is a material return from the pure electronic realm of digital computation. This return does not constitute a mere retreat to hand crafted techniques, instead material model-making techniques, including new industrial processes and robotics, are clenched with immaterial computational explorations in a feedback loop where neither is supposed to be privileged. The biotechnological paradigm allows us to see how these techniques reflect the way organism and environment also involve and evolve simultaneously.

The emphasis on the imbrication of architecture with biological life is what remains the most strident argument formulated by the Emergence and Design Group. It is perhaps no wonder that by the time we get to the end of Michael Hensel's essay, "(Synthetic) Life Architectures: Ramifications of a literal Biological Paradigm for Architectural Design," he wants to reassure his reader that what the experimental design group is proposing is not a modern version of Mary Shelley's *Modern Prometheus*: don't be afraid of Frankenstein's monster (*TTM*, 6, 17). Despite Hensel's qualification, I would like to suggest that questions of monstrosity should not be placed to the side: Tetralogical beings must also be accepted in relation to the plane of immanence, or the plane of nature if designers are serious about investing in such processes. If we heed Arendt's advice then "what is certain is that the measure [of all things] can neither be the driving necessity of biological life and labour nor the instrumentalism of fabrication and usage" (*HC*, 174). Life always exceeds the categories we lay out for it, and the monstrous, for instance, is suggestive of that which exceeds our attempts to order the wild profusion of things that erupt in a world. In Deleuze's late essay, "Immanence: A Life. . . ." our perpetual attempts to account for what life is always fall short. What we can say is that "A life is the immanence of immanence, absolute immanence: it is complete power, complete beatitude."[6] Life is a question ceaselessly posed, a capacity we continue to strive for, everywhere, in all the moments a living subject goes through, sadnesses and joys included: no form, no subject is inherently bad or good, rather, it depends on the compositions they enter into. Furthermore, some compositions can prove to be either more destructive or else more productive than others, depending on the situation at hand.

EXAMPLE THREE: MESNE

Mesne (Tim Schork and Paul Nicholas) is a young Australian firm whose work was exhibited at the Beijing Biennale in 2006.[7] While their speculative projects remain mostly in the realm of the unbuilt their design propositions are nonetheless provocative. Both principals of the firm have undertaken doctoral research in the medium of design at RMIT University and supplement their practice and research with teaching. In 2006, Mesne led an Interior Design Studio from RMIT: a collaborative group project *Screen Resolution*, which resulted in an exhibition at Euroluce, Melbourne (exclusive lighting store). The title, *Screen Resolution*, plays on both our screen mediated existences, as well as the material idea of screens as architectural components. To commence the design exercise, the studio decided on a simple geometrical figure, common in nature, the six-sided, hexagonal honeycomb. The hexagon as base geometrical unit with which the group was going to work was imagined as owning both an electronic and a real manifestation in that it was conceptualised as both organic cell and as digital pixel. The hexagon is also convenient as it immediately facilitates structural efficiency. Each student, or member of the group decided upon a particular fascination to explore. These included color distribution, scaling, lacing, whether the hexagonal cells could be more or less open. These fascinations were then transformed into simple coded rules that were fed through Rhino software, that is, software commonly used for rendering 3D architectural objects. Off-the-shelf rendering software habitually deployed for the purposes of mere representation was adapted, with the guidance and expertise of *Mesne* to specific ends that became integral to the design process.

Schork describes the simple hexagonal cells-pixels as genotypes, and their interaction in the context of a field as the phenotype. It is the language we must listen to, for it is appropriated from biological science, but activated in the field of computer technology and then applied to computational design. Where a genotype describes the internal genetic code and all the inherited traits of a simple unit, for example a gene, a phenotype determines the external explication of this internal code, which becomes manifest in the external characteristics of a form and how these external characteristics are then more or less suited to a given environment. This, in turn, determines the form's survival and effects how the genotype manifests in the phenotype of a form or organism. In much the same way, each participating student can, in the first instance, be imagined as a simple unit that inherits a simple set of rules. When placed in a design environment together the simple units-students collaborate in the neighborhood of each other's concerns toward the creation of a complex system.

Remarkable about this pedagogical-computational experiment is both the outcome and the venturing of new social relations that might be articulated between designers. Together with a 1:1 scaled screen prototype constructed from cardboard, a series of plaster models were printed. The 3D printer (plaster printer or "dust printer" as it is sometimes known), incrementally paints layer upon layer of plaster and watery glue. It is a wet process that integrates the dry part that is the plaster dust. This is also the kind of process that the body artist, Stelarc has speculated upon with respect to the printing of human organs. Much like Lynn's Embryological House, the series of plaster maquettes that were printed constitute an open series to which further models could be added. What is of importance is the processual system that has been configured such that it can be applied in different contexts, depending on the nature of the architectural problem posed. What is more, the process here is not just pertinent to the construction of form, but suggests new social formations among designers, who can come together with what might appear to be disparate interests and create collaboratively a fascinating whole. Still, the motivation behind employing a discourse affiliated with what has come to be known as the new biotechnological paradigm in architecture remains under-theorised.

EXAMPLE FOUR: FOA'S ARK

Foreign Office Architects (FOA) have published a well-known manual of sorts, which they have named *Phylogenesis: FOA's Ark*. The title makes an intentional play on the biblical story of Noah's ark, with the suggestion that all life is held within the confines of its pages: it seems we are asked to read *FOA's Ark* as the promise of all future architectural life on earth. Somewhat like *Morpho-Ecologies*, this publication also operates as a user's guide to the new biotechnological paradigm. The book itself is a neat green package (somewhat like a bible) with a fold-out section that describes a phylogenetic tree of the architectural forms developed by the studio. Phylogenesis is the scientific means of charting a genealogy or line of decent of a living organism, and constitutes a system where evolution and taxonomy come together (*PFA*, 645). The manual reflects on ten years' worth of work (1993–2003), such that a "population" of architectural forms could be gathered that was sufficiently large to apply the organisational regime of phylogenesis. Here FOA share an implicit understanding of the shared compulsion to classify that belongs to biology and architecture alike. Through a taxonomical chart that could be mistaken for a table of living organisms, FOA classify their formal output as though it were a living population, even a "genetic pool" in order to identify the fields of consistency that could be superimposed upon their labour. FOA explain "our practice may be seen as a phylogenetic process in which seeds proliferate in time across different environments" (*PFA*, 8–9). Their specific intention was to create a DNA of their architectural production. In focusing on outward form, façade and orientation, their taxonomical, phylogenetic chart admits a keen interest in architectural skins or surfaces, or that, which faces and filters the external environment.

Across all the examples considered above the question of the responsiveness of the surface conceived as skin, as wall, as built façade insists. Alongside processes that emulate life, the appropriation of genetic algorithms, as well as scientific systems of classification, that which receives increasing attention is the surface, as that responsive membrane which touches our senses. The surface goes deep. The surface is supported by the wetness of live being and the thickness of protection. The surface breathes—even if it is merely mechanical! The surface is the plane that brings live being and architectural material into closest contact, such that zones of indiscernibility are installed between touching and touched. It is possible to begin to imagine that matter speaks back; that matter answers through a language of the senses, having become increasingly intelligent in response to our advancing technological prowess. Or else, we might venture, has matter itself brought us blindly to this point?

Many of the claims of the Emergence and Design Group appear radical, offering transformative potentials for a more sustainable world and an architecture that is living organism. Ingraham explains this potential in terms of the surface effect, "the surface meshes of computational architectures carry the potential not only for acting as some kind of living surface but also for making profound fields of reparation beyond their immediate boundaries" (*AAH*, 29). That is to say, the material management of architecture, digitally augmented, might extend itself through these new technologies to attain more environmentally responsive systems. This is a key aspect of argument forwarded by the Emergence and Design Technology Group, who explain that most form-finding methods result in curved geometries and smoothly differentiated surfaces, as surface curvature allows structural capacity and opportunities for controlling orientation in response to environmental factors (*TTM*, 31). Finally, the curved surface interfaces well with "nature" in more than a merely metaphorical way. What is curious is that there is no necessary connection between the application of biological processes appropriated from nature and what an architectural form looks like. A resulting architecture does not have to look organic or as though it were derived from nature, and yet, so often such a resemblance inheres. To this architectural surface of smooth differentiation what needs to be added are new relationships of intertwinement between human and other kinds of bodies and life forms, the immediate, mostly porous boundaries beyond which the environment and associated pressures insist. As Ingraham intimates, from the apparent rise of a techno-biological paradigm a new metaphysics or perhaps ontology of the surface needs to be articulated. This surface writhes beneath the touch, is animated, suffers peristaltic movements and evolves over time only to pass resolutely away. The theoretical and electronic domains of our computer software/hardware apparatuses show us this process; but are such processual adventures enough? Guattari in *The Three Ecologies* argues for an ecosophy that accounts not just for the environment, but also for social relations and human subjectivity.[8] He argues that an approach to environmental concerns should not forget the co-presence of shifting social relations, the transformative potential of human subjectivity in construction, and relations of subjectification. It should also be noted that Guattari argues for another paradigm altogether, an aesthetic paradigm, which is also a processual paradigm (*CEA*, 106). Importantly this paradigm always responds by way of a double, and asymmetrical surface articulation between infinite speeds of thought as they pertain, on the one side, to a plane of immanence and, on the other side, to the emergence of finite, manifested states of things and bodies. This is what Deleuze and Guattari name the conceptual surface of the plane of immanence, or the transcendental field, which does not describe a transcendent realm, but operates immediately with a present situation, effectively motivating the present, as well as becoming transformed in return.

Two kinds of surface are at work here: the diagram of virtual forces that Deleuze and Guattari illustrate, an active abstract machine that can be injected into the architect's design practices; and the surface that results, and continues to progressively result, unfurling, unfolding, as processes of actualisation find form and make it durable for the meanwhile: topological, processual, digital architecture in its particular instantiation. For Deleuze and Guattari the work of art (and we can locate architecture loosely under this rubric) is defined as sufficiently durable for the meanwhile, sufficient to set into circulation beings of sensation composed of affects and percepts. Architecture is the first of all the arts, Deleuze and Guattari proclaim in *What is Philosophy?* Its key action is to frame a patch of territory, to demarcate inside from outside, and so on (*WP*, 186). In suggesting that the primary function of architecture is to frame, Deleuze and Guattari appear to remain too far away from the breathing wetness of the life, individual, collective, human, animal, and other that comes to participate in the architectural milieu. If we posit instead a continuum of architecture and life, we can also begin to imagine zones of indiscernibility between digital immersive realms and our daily material existences.

The ideology of the imaginary that emerges in contact with the experimentation of digital architecture is marred only where false authority is claimed from the sciences, or from philosophy, and life as a vital force is reified as so many adventurous, adventitious, but empty forms. An empty form is that which answers to no problem, and problems can be relatively serious or frivolous. An empty form answers nothing but its own question, and remains disengaged from the social field, or plane of immanence. If architecture becomes less about built forms and more about an open ended process of continuous variation, participating intimately with the ongoing upsurge of life's interpenetrations, the intermingling of bodies and flights of thoughts, how then do we explore strategies to frame useful segments of the plane of immanence, such that they can be put to use toward pressing contemporary problems?

How do we support the real, immanent intricacy and complexity of social and other relations as they pertain to the aliveness of an architectural milieu? The ethico-aesthetics that I call forth with the title of this essay comes predominantly from the work of psychoanalyst, Guattari, well known for his own work (practical and theoretical), as well as his collaborations with Deleuze. You simply have to open Guattari's book, *Chaosmosis*, to get the sense of the great distances traversed and enormous conceptual and adventurous leaps being made. He tells us "geopolitical configurations are changing at a great pace while the Universes of technoscience, biology, computer technology, telematics and the media further destabilise our mental coordinates on a daily basis" (*CEA*, 119). We are faced with crises that are ecological, social, political, and existential (and it should be noted that Guattari writes this prior to the events of September 11, 2001). Guattari asks, "how do we reinvent social practices that would give back to humanity—if it ever had it—a sense of responsibility not only for its own survival but equally for the future of all life on the planet, for animal and vegetable species, likewise for incorporeal species such as music, the arts, cinema, the relation with time, love and compassion for other, the feeling of fusion at the heart of the cosmos?" (*CEA*, 119–20). This is his ethico-aesthetic challenge to us. And he joins with Deleuze in calling us toward a creative practice, to produce "creative sparks" that touch on all the micro-behaviours of our everyday existences, day in, day out, travelling upon and through the plane of immanence. Guattari recommends that it is less a matter of managing novel cognitive spheres (we don't necessarily require the latest techniques and technologies), rather it is a matter of "apprehending and creating, in pathic modes, mutant existential virtualities" (*CEA*, 120). How then do we make ourselves worthy of what we are fully capable? If we remember the ever mobile, undulating surface of the plane of immanence, we are asked to imagine how transversal or non-hierarchical relations are created that might augment rather than diminish a capacity of existence through conceptual and sensory becomings of all imaginable kinds. Guattari refers to this creative activity, this ethico-aesthetics, as a cartography. We are all mapmakers, drawing up diagrams of sense on a daily basis and our milieu is the plane of immanence.

Now the problem with architects is that in becoming preoccupied with novel distributions of matter and form, they do not always remember to conjoin thought and matter toward the framing of pertinent problems. It is always important to stress that for Deleuze and Guattari, concept construction must secure itself to contemporary problems, without which the concept would wander, disengaged, as empty form.

The plane of immanence is also like a plan awaiting our experiments of actualisation, that is to say, "The map expresses the identity of the journey and what one journeys through. It merges with its object, when the object itself is movement."[9] But architects too often forget the constellation of affects and percepts that attend the plane of immanence, that one can really become transformed, depending on what encounters one has. These are our sensory and conceptual becomings, whereby we become something other than what we once were. Instead, the digital or computationally derived building-object too often emits novel signs and gizmos, technical prowess and innovative techniques, and forgets the life that moves in its midst, forgets that it is already part and parcel of this life.

How is architecture then taken up in the relations that articulate the plane of immanence? Architecture, like art, is an incorporeal species of being as well as a coagulator of actualised material effects. The work of art and architecture too, for those who use it, is "an activity of unframing, of rupturing the surface, of baroque proliferation or extreme impoverishment, which leads to a recreation and a reinvention of the subject itself" (*CEA*, 131). The main battle to be fought is with habit, cliché, and opinion and often architects might ask themselves, how little design do I need to undertake?[10]

NOTES

1. In her book *Biomimicry*, Janine Benyus describes the close vision of the oyster shell; a University of Washington research group investigates nacre, or oyster, and abalone shell: "the intricate crystal architecture" is composed in cross-section of hexagonal disks of calcium carbonate (chalk), stacked in a brick wall arrangement, but importantly, between these hard bricks, a mortar of "squishy polymer" allows for stress to be accommodated like a ligament. According to Benyus's account of what the researchers have discovered, the shell deforms under stress and behaves like a metal. Thus, the wet and the dry come together to create the renowned hardness of the shell. Janine M. Benyus, *Biomimicry: Innovation Inspired by Nature* (New York: Harper Collins, 1997), 98–9.

2. See Diane Agrest for a treatment of the issue of what she calls transsexual (and what I prefer to call transgender) architecture. Agrest argues that this is present throughout the history of architecture in the propensity of an architect to take on the metaphorical roles of both mother and father in the "conception" and "birth" of their built forms, which effectuates the appropriation and erasure of the woman's body. Diane Agrest, "Architecture from Without: Body, Logic and Sex" in *Assemblage* 7 (1988): 29–41.

3. Greg Lynn, "Architectural Curvilinearity: The Folded, the Pliant and the Supple," in *Folding in Architecture*, *AD*, revised edition (2004): 24–31.

4. Michael Hensel, Achim Menges and Michael Weinstock, eds, *Emergence: Morphogenetic Design Strategies*, *AD* 74, no. 3 (May/June, 2004). Hereafter cited as *EMD*. Michael Hensel, Achim Menges, and Michael Weinstock, eds., *Techniques and Technologies in Morphogenetic Design*, *AD* 76, no. 2 (2006). Hereafter cited as *TTM*.

5. Michael Hensel and Achim Menges, *Morpho-Ecologies* (London: Architectural Association, 2006).

6. I have adapted the translation by using the term beatitude. Boyman has translated the original term "beatitude" with "bliss." Gilles Deleuze, "Immanence: A Life. . . ." in *Pure Immanence: Essays on a Life*, trans. Anne Boyman (New York: Zone Books, 2001), 27.

7. See http://www.mesne.net/.

8. Felix Guattari, *The Three Ecologies*, trans. Ian Pinder and Paul Sutton (London: Athlone Press, 2000), 28.

9. Gilles Deleuze, "What Children Say," in *Essays Critical and Clinical*, trans. Daniel W. Smith and Michael A. Greco (London: Verso, 1998), 61.

10. A version of this essay entitled, *In Search of an Ethico-Aesthetics for Wet Architecture*, was first presented as a keynote for the conference *Ideology of the Imaginary in the 21st Century* held at the Experimental Arts Foundation, Adelaide, South Australia, March 1-2, 2007. I would like to thank the organizers, especially Teri Hoskin, for the invitation to present an earlier version of this paper, and for their invaluable feedback. I would also like to thank Ken Bolton for his kind feedback and editorial suggestions. An earlier version of the paper is also is available online at http://eaf.asn.au/2007/symposium_p_frichot.html. The paper was subsequently developed to frame a panel at the IAPL 2008 Conference, Global Arts/Local Knowledge, which is now presented in part III of *De-signing Design*.

Chapter Eleven

Digital Organic Design

Architecture, the New Biology, and the Knowledge Economy

Karen Burns

Once again architecture, in search of its lost object, is contaminated by this model fever. Christopher Alexander is already a precursor, and models can now be seen everywhere. They become the architectural avant-garde, bringing a kind of scientific guarantee given the tool of mathematics (which through science in its own domain become techniques when applied elsewhere—a phenomenon little understood by those who believe in a sort of osmosis whereby architecture, through the application of mathematical models, can itself become a science) (*AWT*, 76).

Knowledge has become an economic phenomenon. As one economist, Dominique Foray argues, since the 1970s, new economic formations have emerged: knowledge-based economies, defined by the proportion of "knowledge-intensive jobs."[1] Foray observes that "science and technology tend to be central to the new sectors tending to give momentum to the upward growth of the economy," and that these realignments "are reflected in an ever-increasing proliferation of jobs in the production, processing, and transfer of knowledge and information" (*EK*, ix–x). Over the last fifteen years, architecture's engagement with the disciplines of science and technology parallels this broader historical transformation of post-industrial societies. The economic calibration of knowledge, its "economic characteristics," and status as a "good" and the financial valuation of knowledge transfer and reproduction, are not addressed by this paper. However, I wanted to mark the origins of the term "knowledge transfer" in the discipline of economics, because this paper is concerned with one architectural case of knowledge transfer and the complexities attending this move, where the economic term "transfer" becomes somewhat inadequate to the task of describing the reformation of scientific claims within the discipline of architecture. This chapter investigates one architectural transformation of evolutionary theory and argues that the reformulation and rewriting of material extraneous to architecture involves another technology, that of architecture.

Greg Lynn's Embryological House Project (2000) is one of a number of widely circulated contemporary projects that mark architecture's intersection with the specialised discourses of biology.[2] The text accompanying the publication of this project forms a central document for this chapter. The presence of two specialist disciplines, biology and architecture, and the intriguing question of their intersection are staged in an impressively seamless movement in the opening paragraphs. After a series of digital renderings of his design, and captions such as embryo and egg—seeming to signpost the place of biology in this project—Lynn's text opens with architectural claims about new, contemporary modes of production and aesthetics:

> The Embryologic Houses can be described as a strategy for the invention of domestic space that engages contemporary issues of brand identity and variation, customisation and continuity, flexible manufacturing and assembly and, most importantly, an unapologetic investment in the contemporary beauty and voluptuous aesthetics of undulating surfaces rendered vividly in iridescent and opalescent colours (*CPA*, 31).

This detour from biology via traditional architectural concerns marks the interface of two discourses, and the project of reworking one via the other. The point of intersection begins to be clarified in the sentence: "The Embryologic Houses employ a rigorous system of geometrical limits that liberate models of endless variations" (*CPA*, 31). Addressing brand identity and variation allows "recognition and novelty" and "design innovation and experimentation" (*CPA*, 31). All of the implications of this form of production, which Mario Carpo terms "non-standard seriality," "mass producing a series in which all items are different" will not really concern us here, but of interest is the deployment of economic terms from late capitalist modes of production to form the links between discourses.[3] The final part of Lynn's first paragraph provides the next linkage in the chain. The chain has so far moved from economies of production/consumption, to an aesthetic claim, to design techniques, back to avant-garde aesthetic terms (innovation and experimentation) and finally a larger picture emerges in this last sentence:

> In addition to both design innovation and experimentation, many of the variations in the Embryologic houses come from an adaptation to contingencies of lifestyle, site, climate, construction methods, materials, spatial effects, functional needs and special aesthetic affects (*CPA*, 31).

The word "adaptation" is possibly drawn from biological discourse and this connection seems more substantiated by the next paragraph, which begins, "There is no ideal or original Embryologic House. Everyone is perfect in its mutations." Moreover the "formal perfection derives from "a combination of the unique, intricate variations of each instance and the continuous similarity of its relatives." And then, after indicating that the variation occurs in the relationship between the generic envelope and a fixed collection of elements, Lynn delivers his final sentence of the second paragraph and makes a larger historical claim, "This marks a shift from a modernist, mechanical technique to a more vital, evolving, biological model of embryological design and construction" (*CPA*, 31). Here borrowings from the discourse of biology are marshalled to produce a new internal history of architecture. This is one of the strategic effects of citing biological discourse. It shapes a certain mode of contemporary architecture as a more naturalistic mode of production. The place of a new economic formation, "mass customisation," is eclipsed by the realignment of the new "biological" mode within a longer architectural history premised on a binary formulation: of older mechanistic versus new biological paradigms.

The appearance of words normally exterior to the discipline of architecture—adaptation, mutation, relatives, and of course embryology—all of which are biological terms, raises the intriguing issue of the strategic effect of these citations in an architectural discourse. The first three terms in particular are closely associated with evolutionary theory. The next part of my chapter involves a close analysis of the disciplinary outlines of evolutionary theory in order to investigate the status and meaning of the scientific discipline's particular terms when they are displaced onto architecture.

Evolutionary theory seeks to account for a particular kind of biological change: variation, the ways in which variations in organisms give rise to new species, the ways in which those variations are transmitted over generations, the mechanisms of heredity, how these variations are "selected," that is, survive, the belief that some of these variations may be beneficial, and that there is a correlation between variation, adaptability, and survival, demonstrating that adaptation ensures greater survival. The field is vast, specialized and complex, and most importantly, full of disagreement, hesitations, qualifications, and uncertainty. These contests mark the place of evolutionary theory as a social discipline, comprised of competing or different accounts. Some of these disagreements can be recounted by exploring the complexity of terms such as mutability and variation, two of Lynn's key terms.

Evolution is in one sense a biological version of history. It seeks to account for change. Transformation, difference, and the persistence of certain transformations, their triumph is viewed and noted. Evolution relies on a model of temporality, like history, to understand and judge its material. It operates with a notion of inheritance, the traits transmitted from generation to generation. These qualities and their persistence can only be known retrospectively. Only by looking back can scientists decide which traits and behaviors have been transmitted and selected over time. There are many debates as to whether this is a slow process that is gradual—very, very gradual—or whether there can be rapid genomic restructuring.[4] (And it is not clear to me what rapid might be in terms of evolutionary time). Moreover, the problem of what constitutes the targets of selection—genes or individuals, groups or species—has been debated, most adamantly by Richard Dawkins and Stephen Jay Gould (*EFD*, 38). Moreover, it is possible that the evolutionary process may be entirely random and any historical model premised on causality and determinism might fail due to the operations of contingency. In other words, individual agents—at the level of individuals, groups or species—may play no part in the persistence and reproduction of survivable traits.

Variation is complex and entails several possible mechanisms. Heritable variation occurs through genetic mutation and also sexual reproduction. Mutation, with which Greg Lynn is concerned, refers to changes in DNA sequences. The reasons for these changes are variable caused by internal imperfections in the copying process, by other internal activities, or by external causes. However, mutation is not considered to be a primary factor in variation. Mutation rates are deemed to be low, because lineages with good heredity needed faithful transmission dependent upon accurate copies of genes.

The second form of variation, one that Lynn does not address, although it is considered to be the primary cause of difference, is sexual reproduction. Sexual reproduction produces enormous variation and is the most obvious source of genetic variation. Offspring are never equally mixed and equally weighted clones of their parental material. The importance of sexual reproduction as the most obvious source of variation was skewed early in the last century when a number of theorists, such as Hugo de Vries and William Bateson had argued that evolution occurred in big leaps. For de Vries, "the driving force in evolution was mutation, a process that suddenly and without cause irreversibly changed the germ plasm (a part of the chromosomal material set aside for eggs, etc, whatever gives rise to the next generation). Mutation "produced a new type of organism in a single step" (*EFD*, 23). This thesis remains highly controversial. Mutations are new genetic variants but in evolutionary terms, their importance is always measured within a longer time span. Will the mutation survive into the next generation and will it be selected?

I have spent some time outlining some of the major disagreements in evolutionary theory in order to establish the ways in which major terms and theories remain under contest in this expert discipline. These quite different investigations of key terms introduce a number of levels of complexity in the problem of accounting for cause and effect in evolutionary change. Terms that I had assumed were stable, become much more complex due to the range of possible explanations. These disagreements are not noted in Lynn's discourse, and through this omission, key terms destabilised in evolutionary theory become much more stable and certain when deployed in an architectural setting. Later, I will address the issues generated by this transformation; the problem of how we should read such specialised technical terms when they are radically disjoined from their former expert domain.

In part, I have given this non-architectural account of the contested nature of key terms in the discipline of evolutionary theory and of evolutionary theory's key mechanisms because I am interested in marking the radical incommensurability of parts of evolutionary discourse with architectural modes of production. I note this disjunction in order to later address the problem of how we should read the architectural use of evolutionary theory when architecture cannot fulfil some of the key criteria of evolutionary discourse. Two dissonant architectural areas require attention because of their strident deviation from the original scientific discourse. One is the limited definition of evolutionary variation in architecture and how variation operates, for instance, Lynn's focus on mutation not sexual reproduction, and the other domain entails the difficulty of imagining how the evolutionary selection mechanism would operate in architecture.

Since computer software simulation programmes do not have the biological capability to breed and reproduce, it is understandable that Lynn would focus on mutation rather than sexual reproduction. Mutation however, creates new variations in genes, within one reproductive cycle. It offers a shorter time span. Mutation engages directly with the problem of iteration as a copying process, since mutation is a differential process in copying material. However, as noted above, in current evolutionary theory, mutation rates in lineages that survive are deemed to be low. So while mutation occurs it is disjoined from evolutionary success.

Another problem occurs when evaluating variation in architecture due to the production cycle of design. Evolutionary history, imagined here through the mechanism of selection across generations, is the only way of measuring transmission and survival of variations, no matter their source. Variations need to be heritable across generations. Even if we take the time between human generations to be sixteen years, it is in no way equitable to the temporal dimension of computer iterations.

Perhaps we are talking about fruit flies or E. coli bacteria with shorter time spans. I am presuming because of the title of the project "embryological" and its morphology that we are referring at the very least to a mammalian embryo. The non-correlation of evolutionary time and design or production time remains problematic in this discourse.

Even setting the issue of temporality aside, another problem persists: the selection mechanism. Evolutionary history is a form of history written for victors. There may be many contingent factors that ensure the survivability of certain traits over others. Success in this endeavour can only be known and judged after the fact, never in the midst of the event. Given that selection operates as a mechanism outside and above individuals, it may never be able to be harnessed and determined by them, and certainly not in their lifetimes since it must be transmitted and evaluated across generations. In other words, any architect or generation would have to leave the evaluation of their work to a historical process. Only the long span of time confers success and legitimacy on the project's claims to adaptability and mutation as a form of success. Otherwise any architectural project could just be a mutation that has no benefit or success in evolutionary terms. It could just be one mutation among many.

I have noted three effects in this operation of "knowledge transfer" of evolutionary theory into architectural discourse: the production of a new internal history of architecture, the selected deployment of terms associated with a scientific discipline to produce new modes of description of architectural production, and the production of a certain stability around terms that are unstable and contested in their original scientific domain. Moreover, I have suggested that a radical incommensurability prevents us from using evolutionary theory to evaluate current modes of architecture in evolutionary terms.

Attempting to read architecture's use of evolutionary theory as extensions of a scientific, technically expert discourse has produced a certain number of difficulties. Architectural design and production is not an extension of evolutionary theory but a distinct discipline. Even when architecture shares similar techniques with scientific fields—such as data modelling techniques used to model flows of weather data or the mapping of molecular energy landscapes, techniques which have been discerned in Greg Lynn's processes—the displacement from original fields of use generates intriguing differences (*ASE*, 347). My concern is with these differences and the status of these distinctions.

I will confine my discussion, for the sake of brevity, to the function of language in marking these differences. I have focused on the discontinuity of meaning in the appearance of terms generated by one discourse when deployed in another. If the terms of evolutionary theory, which erupt in architecture do not achieve the complexity of expert, technical scientific discourse what are the reading conditions that govern our understanding of these words in architecture? I will argue that these terms function metaphorically.

Susan Sontag, in the opening paragraph of her book *Aids and its Metaphors*, quotes Aristotle's work *Poetics* to offer a succinct definition of metaphor, "Metaphor consists in giving the thing a name that belongs to something else (*AM*, 5). This denotes the ways in which metaphors trade in the traffic between resemblance and difference. Aristotle's use of the term occurred in a text on literature; Sontag's book reminds us of the migrations of metaphor from a specialised tool of literature and its studies of figurative language into a form of analysis of ordinary language and technical languages, occurring within many disciplinary domains in the twentieth century. The role of metaphor as a component of non-literary language was inaugurated by the work of early twentieth-century linguists such as Roman Jakobson in his study of folktales. In the later twentieth century, linguists and anthropologists such as George Lakoff and Mark Turner increasingly focused on the role of metaphors in so called ordinary language. Apart from Sontag's study of metaphors in certain medical conceptualizations of illness such as cancer and AIDS, a number of philosophers and historians of science (Lilly Kay, Mary Jacobus, Evelyn Fox Keller, and Nancy Tuana) have studied the role of metaphors in conceptualizing science, in particular biological discourses. These studies examine metaphor in order to understand the ideological function of knowledge formation. But they also suggest the ways in which shifts in a discipline's knowledge domain are given shape by new metaphors.

Aristotle's attractively brief description of the figurative function of metaphor should be supplemented by the definition it has acquired since the later 1970s. In their study *More Than Cool Reason*, George Lakoff and Mark Johnson observe, "a metaphor is not a linguistic expression, it is a mapping from one conceptual domain to another" (*MTC*, 203). This account usefully describes the appearance of terms from evolutionary theory in the discourse of architecture. A metaphor generated from the importation across disciplinary borders provides a shorthand way of grasping a relationship between apparently dissimilar discourses or practices. Deploying metaphors is a compressed, shorthand mode of communication, and a way of producing a new proximity between geographically distant and conceptually dissonant material.

Over the last twenty years architectural historians and theorists have investigated the creation of analogies to describe the traffic between architecture and other disciplines. A number of writers have analyzed architecture's distinctive use of material from fields exterior to itself, most particularly, the relationship between philosophy and architecture.[5] Studies by Catherine Ingraham and Mark Wigley attempted to examine how architecture functions metaphorically for other disciplines. Ingraham argued that architecture operated by force of its metaphoric status in culture, apparently designating the proper forms of inhabiting space; and Wigley examined the functioning of architectural terms such as foundation within philosophy, a discourse in which architecture was mobilized to ground philosophy's authority claims.

A more recent architectural study, Adrian Forty's *Words and Buildings*, presents a detailed analysis of both language and scientific metaphors within the history of architecture. Forty traces the emergence of certain metaphors within architecture and sometimes evaluates the historical success of particular metaphors. His definition of metaphor concurs with that offered by Lakoff and Johnson, the "characteristic of an effective metaphor is that it borrows an image from one schema of ideas and applies it to another, previously unrelated schema."[6] Significantly he observes, "metaphors are never more than partial descriptions of the phenomena they seek to describe . . . indeed were they to succeed in total reproduction they would cease to be metaphors which subsist through likeness drawn between inherently unlike things" (*WB*, 84). Once material has left its original disciplinary field, such as evolutionary biology, there is always the possibility that it will start to operate as a metaphor, a point of resemblance and as a substitute for the discourse it has left behind. In fact, this is precisely Ingraham and Wigley's argument about the metaphorical status and power of architecture in culture and in philosophy.

Architecture, of course, is not unique in transforming material extraneous to its discipline into metaphors. However, architecture provides a spatial formation and realisation of these alignments. It gives evolutionary theory a spatial imaginary, and one that is distinct from, although proximate to, the uptake of evolutionary theory into economic and managerial business models of late capitalism, a project externalised by the founding of the *Journal of Evolutionary Economics* in 1991. The discipline of architecture's capacity for spatial realisation marks the distinctive work of architecture in moments of knowledge transfer, as particular disciplinary domains are reformulated in crossing the border into architecture.

This paper was originally written in response to the Australasian Architecture Schools' 2007 Conference proposal that knowledge transfer "threatens the consistency and specificity of architectural techniques." I would argue that there is always an ever-present technology of architecture that converts material into spatial realisations, and realigns external material into forms of knowledge interior to the discipline of architecture. These operations could be usefully described as a "technology" of architecture, reworking Michel Foucault's famous observations on the "technology of sex" as a set of techniques.[7] Foucault invented the term for strategic ends, in order to disrupt normative definitions of technology as inventions and techniques. Deploying the category of technology outside its usual domains, Foucault enlists his newly rewritten term in order to denaturalise one phenomenon: sex. He mobilizes technology to designate the systematic techniques organizing a field of knowledge, even one which appears biological and thus natural. He redefines the etymology of technology in order to analyze knowledge formation. Arguing that intellectual domains are determined by structural rules and techniques determining what counts as knowledge, Foucault demonstrates that by delineating a terrain a discipline controls the form in which questions can be asked and thus what can be asked at a given historical moment. A discipline is not necessarily marked by the sum of its internal knowledge, but by its operations, "The 'economy' of discourses—their intrinsic technology, the necessities of their operation, the tactics they employ, the effects of power which underlie them and which they transmit—this, and not a system of representations, is what determines the essential features of what they have to say" (*HS*, 68–69). Technology denotes the apparatus organizing knowledge formation and insistently marks the nexus of knowledge and power.

Architecture rewrites terms and fragments from scientific discourses into strange mutations. These transformations mark architecture's inside; a terrain where external ideas are not merely imported but formulate new internal histories and theories in architecture, where the technology of architecture realigns material into its own disciplinary formations. Older architectural terms and questions are both continued and discontinued in this formation. Evolutionary theory offers a model for investigating notions of generation without a human operator (autogenesis), the rearticulation of temporal rupture as a mode of innovation (an avant-garde investment in the new) and the use of evolutionary theory as a model of history to establish legitimation via the historical validation of adaptation, selection and survival. Biology offers an ecological model of the environment imagined in network and information terms. The "organism" or embryo offers a source for form generation. This "Nature" would almost naturalise the workings of ideology, producing a transparent and readable nature, different to that posited by one philosopher of science who describes the "mystifying and recalcitrant chaos of higher level organisms."[8] Architectural processes are modes of projection, of transference as well as transfer. As methods they inscribe the force of human editing, selection, and rewriting of material. These social operations form a discourse, ensuring that its tactics and modes of legitimation are all too human, even if its surface may appear otherwise.

If the terms of evolutionary theory are evacuated of their technical complexity when deployed in Lynn's architectural articulation, this does not make them uninteresting or un-useful. Metaphors can mark the place of a complex process of creative appropriation. These tropes of figurative language may well be the starting point for a process of creative generation. They image a new relationship between apparently dissonant materials. They visualise an idea or operation and make it known in the first place so that it can be further investigated and provide the primary point of creative work. However, a problem arises, not in the metaphors themselves but in the work they are called upon to undertake. In the quotations from Lynn cited earlier in this essay we can see how easily metaphor sheds itself of its primary operation of bringing into proximity two unlikely categories or schema, to become instead, a statement of identity. In this way of course, metaphors are more rhetorically commanding because they do not contain modifiers, such as we find in the definition of a close relation to the metaphor, the simile. As M. H. Abrams observed many years ago, a simile "is a comparison between two distinctly different things (is) indicated by the word 'like' or 'as.'" The modifier in a simile reminds us of the uneasy oscillation between different disciplinary territories or nouns and abstract nouns.

While it might appear to be mere literary pedantry to note the similar operations but differing rhetorical forms of metaphors and similes, key questions of authority and authorisation underwrite these distinctions. In the realm of technical expert usage, key terms of evolutionary theory might be contested but they function as descriptions, not metaphors. Architecture's figurative use of these terms becomes problematic when the metaphor masks or forgets the operation of comparison. A metaphor can perform what a simile bound by its stated form of comparison has more trouble enacting: collapsing the difference between the claims of creative usage and making powerful authority claims. In denoting the difference between creative and authority claims, few contemporary architectural commentators are as scrupulous as John Frazer in his 1995 book, *An Evolutionary Architecture*. Early in the text, he distinguishes between a scientific hypothesis and a design hypothesis and he insists on the nature of inspiration.[9] Even a distinguished and careful critic such as Mario Carpo, in a recent essay theorising Lynn and Bernard Cache's use of software simulated designs to form a variable set, a "non-standard series," slips effortlessly into a problematic identity statement, remarking on an "algorithmically defined fixed genera and endlessly morphing species" (*L*, 106). For all of the reasons I have argued above, this evolutionary metaphor gives striking form to a new idea, but does not bear the weight of close scrutiny as a description verifiable by scientific evidence, since architectural design production fails to fulfil the criteria of evolutionary theory. However, as a metaphor, an applied borrowing from one conceptual schema onto another conceptual schema, it denotes the production of a new relationship to produce different knowledge within our discipline; in Carpo's example, to rewrite models of authorship and aesthetic criteria. Moreover, metaphors have rhetorical force because they function figuratively; they offer a striking image, a visualization of an idea, and their effect can be ascertained by comparing the differences between an abstract formulation "mass producing a series in which all items are different" and the fixed genera, endlessly morphing species metaphor deployed by Carpo. I can remember the latter phrase and visualize it, but not the former.

The slide between the use of metaphor to produce creative analogies and the use of a metaphor to ground authority claims lies at the center of the discipline of architecture. Deterministic and authoritative accounts of design, rather than acknowledgements of creative appropriations prevail in architecture. The remarkable appropriation of contradictory or contested and difficult theoretical material into compressed syntheses and useable models is an extraordinarily creative process, but by no means is it logical or inevitable. The process evades scientific authority claims but makes the claim of creative authority. I do not see this as problematic, unless we fail to make the distinction. Perhaps the use of metaphors is intensely involved in questions of authority, if we remember Mark Wigley's analysis of the function of architectural metaphors, such as foundation in the discipline of philosophy. Here architecture is invoked to ground philosophy's authority claims.

A larger project might further extend the symbolic significance of evolutionary theory in contemporary architecture in order to investigate the historical conditions surrounding this kind of "knowledge transfer." Michel Le Dœuff in a study of imagery in philosophical discourse argues that the "meaning conveyed by images works both for and against the system which deploys them."[10] Functioning as points of tension and sometimes contradiction, images can "sustain something which the system itself cannot itself justify, but which is nevertheless needed for its proper working" (*PI*, 3). As an architectural historian, I would speculate that the use of evolutionary biology metaphors not only demarcate an outside to architecture, but allude to a larger exterior context, one that supports our work, but is invisible in naturalised presentations of evolution: the various alignments of evolutionary theory and science and knowledge within the complex political and social formations of post-industrial capitalism. This is a subject for another essay.

NOTES

1. Dominique Foray, *The Economics of Knowledge* (Cambridge, MA/London: MIT Press), ix. Hereafter cited as *EK*.
2. Greg Lynn, "Greg Lynn: Embryological Houses," in *Contemporary Processes in Architecture*, *AD* 70, no. 3 (2000): 26–35. Hereafter cited as *CPA*.
3. Mario Carpo, "Tempest in a Teapot," *Log* 6 (2005): 99. Hereafter cited as *L*.
4. Eve Jablonka and Marion J. Lamb, *Evolution in Four Dimensions* (Cambridge, MA: MIT Press, 2005), 70–71. Hereafter cited as *EFD*.
5. Catherine Ingraham, "The Faults of Architecture: Troping the Proper," *Assemblage* 7 (1988): 7–13; Catherine Ingraham, "Animals 2: the problem of Distinction," *Assemblage* 14 (1991): 25–29; Michael Speaks, "Ti's Out There. . . . The Formal limits of the American Avant- Garde," in *Hypersurface Architecture*, *AD* 68, no.5/6 (1998): 26–31; and Mark Wigley, *The Architecture of Deconstruction* (Cambridge, MA/London: MIT Press, 1993).
6. Adrian Forty, *Words and Buildings: A Vocabulary of Modern Architecture* (New York: Thames & Hudson, 2000), 100. Hereafter cited as *WB*.
7. Michel Foucault, *The History of Sexuality Vol 1*, trans. Robert Hurley (Harmondsworth: Penguin, 1990 rpt), 90. Hereafter cited as *HS*.
8. Evelyn Fox Keller, *Refiguring Life: Metaphors of Twentieth-Century Biology* (New York: Columbia University Press, 1995), 81.

9. John Frazer, *An Evolutionary Architecture* (London: Architectural Association, 1995), 12.

10. Michèle Le Dœuff, *The Philosophical Imaginary*, trans. Colin Gordon (London: Athlone Press, 1989), 3. Hereafter cited as *PI*.

Chapter Twelve

De-signing as Bio-Technological Endosymbiosis

Stephen Loo

The appropriation of the biological paradigm in contemporary digital design practices frequently draws on the biophilosophy of Gilles Deleuze and Félix Guattari. These particular concepts lie at the nexus of organismic bodies, internal and external energetics, and systemic (self)actualizations or individuations, concepts that frequently go under the theoretical ambit of "becoming."

What Deleuze and Guattari's biophilosophy provides is a model of genesis that radicalises the design process in general; and this is especially so for digital design practices. Take for example the appropriation of concepts such as "body-without-organs" (BwO) in design, which brings with it an understanding of the virtual dimension in design that wrests the meaning of the "virtual" away from the commonplace signification of immaterial digitally-mediated worlds, towards a condition that describes entities (whether material or digital or conceptual) poised in potentiality. Therefore, within digital design such thinking holds significant implications for rethinking design. This is because it makes contingent the very disciplinary definition of the field. It does this by evoking a radical contingency between three realms: the informational non-physical realm, which is the conventional milieu of digital design practice; the realm of potentiality, which is in and of itself the realm of the non-actualized; and the realm of the physical real, which design is ultimately committed to in its labor.

THE TECHNICITY OF A BIOLOGICAL MODEL

Why has a biophilosophical concept like BwO become a key model for a contemporary understanding of digital design? In BwO, Deleuze and Guattari posit the possibility of a body or bodies outside any determinate space, stratification or identification, ready for any action in their repertories, whose next steps are determined only by virtue of their internal organization. Such a theorization of becoming therefore emphasizes a systematic correlation between entities of various orders, scales and magnitudes at the conjunction of the event; whereupon a body is seen from the point of view of its potential or virtuality, and not as something actualized. By emphasizing the systemic correlation of entities, Deleuze and Guattari make disjunctive all entities that fall upon the plane of existence, but whose presences on that plane manifests the plane itself as providing the synthetic constitution of the entities that is not closed by figuration. A concept like BwO "molecularises" is what makes up existence. This biological paradigm sees design objects, their interactions with and value to human life, even perhaps life itself, not as the sum interaction of individualised entities, but as relations between singularities, relations that cannot be pre-empted, or pre-categorized by their identities.

The biological egg is Deleuze and Guattari's BwO *par excellence*, whose intensive entities contain limitless possibilities for effectuation. This potential for effectuation does not diminish upon actualisation or morphogenesis (when the egg becomes an embryo-body). It is maintained as potential and immanent to the form of the effectuated organism, as something that continues to organize subsequent movements of becoming.[1] What this means is that when applied to the macro conditions of design, Deleuze and Guattari's model of molecular life, which is only ever in a state of constant becoming, operates on a plane where the conventional stratifications and territories that govern how life appears—namely the boundaries between bodies and the environment, human and machine, biology and technics, and so forth—have to be seen as transversal or flexible. "Becoming" involves the pure relationality between entities, and takes place "in-between" concepts, bodies and environments; whereby boundaries are continuously crossed and reconfigured. Molecular becoming effectuates life in what Manuel DeLanda calls a "non-organic" sense,[2] because life is inhabited by a multiplicity of self-organising processes amenable to mathematical and physical expression. The emphasis of this model of life that has so much influenced current digital design technology, is on the differential rhythms and varying levels of intensities of potential actualisations that can be mapped and navigated, rather than the actual entities in their relations; that is, the potential of the process and the emergence of potentialities rather than actual designs or the individuality of their human creator.

It seems that while a concept such as BwO is inspired biologically or embryologically, it nevertheless carries with it the comportment of a physical mechanism, where the potentiality of molecular becoming resides in the systematicity of physical organizations, or using a Deleuzo-Guattarian term, "abstract machines" that consist of "*unformed matters and nonformal functions*."[3] Therefore, the philosophers' biological object is not so much an evolutionary, but a machinic assemblage, working in adjacency to the other equally machinic assemblages, whether social, economic and technological. All such assemblages therefore continuously deterritorialize and reterritorialize relations between the biological and the technological. As Deleuze and Guattari say, "[t]here is no biosphere or noosphere, but everywhere the same Mechanosphere" where "microphysical sectors can serve as an immediate substratum for organic phenomena."[4] DeLanda extends this condition by introducing the idea of a "machinic phylum" as a grouping of attractors and bifurcators in various strata that constitute the source of variability and therefore creativity of the systems in question. The biological model can now genealogically account for organic and non-organic, living and non-living entities alike, as the machinic phylum "designates a single phylogenetic line cutting through all matter, 'living' or 'nonliving,' a single source of spontaneous order for all of reality" (*ZI*, 138).

It is the implicit technicity of molecular machinic assemblages within this biological model that allows digital design, with its implicit desire to displace authorial expression from the production of the object, to dissolve the boundaries between the designer as an individuated entity and the environment with all its physical systems.

And, interestingly, it is also this technicity of the biological appropriations that allows Deleuze and Guattari to posit a particular consistency of entities within their theory of becoming as that, which operates in a *molecular* domain with autonomous possibilities of evolution. I hope to point in this chapter to several shortcomings of Deleuze and Guattari's project to use biology in this way to analyze, intervene, and celebrate the self-ordering micropolitics of contemporary systems. Because much of the uptake of the biological paradigm in contemporary digital design has been influenced by such Deleuzo-Guattarian theories of becoming, this essay also points to the shortfalls of shifting the understanding of genesis in digital design processes wholly away from extensities—the outside conditions of already individuated entities and effectuated conditions which includes the power (and politics) of pre-established forms, namely technological machines, computing softwares with their imaging limitations, the designer as a human individual, and so forth—toward an autonomous domain of preindividual singularities and molecular becomings.

QUESTIONS CONCERNING TECHNOLOGY: RECENT (BIOLOGICAL) DEVELOPMENTS IN DIGITAL DESIGN

Michael Hensel and Achim Menges, in *Versatility and Vicissitude: Performance in Morpho-Ecological Design*—their most recent guest-edition of the influential *AD* journal, and arguably a barometer of the latest trends and developments in architectural practice—call for "performance-oriented design processes that require novel skills and methods to achieve synthesis of versatility and vicissitude" as alternatives to prevailing approaches to sustainability.[5] This performative approach relies upon a versatile and dynamic attitude to the process of design, that is, less a focus on the object of design, and more on its behavior within a specific context. Hensel's and Menges' use of the term "vicissitude" is central to ecology and their concept of "morpho-ecological" design, which they take to mean the variation and mutability in nature or life entailing the "differentiation of the object and the dynamic of the environment" (*VV*, 7). Variability in nature and organisms, and equally in design artifacts and processes, as discussed in this issue of *AD*, is the key link to the authors' initial excursions into new approaches to architectural design as documented in earlier issues of *AD*, namely *Techniques and Technologies in Morphogenetic Design* (2006) and *Emergence: Morphogenetic Design Strategies* (2004). It is in these two journal issues that Hensel and Menges postulate what seems to be an ontological connection between the biological and architectural design processes.

The appeal of a biological paradigm in architecture is, in short, to find relations between the material world and the possibility of its own generation, development, and therefore sustainability. Rather than the human being as creative source, the biological paradigm instates the non-human world with an interior, perhaps primordial, creative force given by a certain consistency or organization of the beings within it. Through methodologies such as biomimicry, cellular automata, material aggregation, and self-organization, the research projects reported by Hensel and Menges enact a discursive continuity towards an ontological consistency between the behavior of organisms and that of technologically mediated productions. The main criterion for the transfer of the biological paradigm into architecture is the capacity for self-organization and dynamics. The biological is made equivocal to transformations of states and material—a commonplace example is the capacity of metabolism—to communicate and transfer information, control internal states in relation to the external environment, development, and reproduction, and finally, mortality.[6]

I find that the treatment of the biology as the impetus for recent developments in design do not differ radically from those in the 1960s and 1970s. Three decades ago there was a belief in the ability of a techno-rationalist framework to colonize biological processes towards an anthropomorphism—

human-centeredness—thus making the processes amenable to human understanding and control. Today, developments in digital computing provide the high level computation required to represent, visualize, and process the combinatory and variational complexities of biological behavior in order to posit equivalences in material and structural behavior. Biology is seen as being the result of the functioning of a group of interlinking parts that provided the dimensions of interaction between the environment and the human agent, and thus the operational possibilities—a "vitalism" or life so to speak—of a creative being. The aim is to look for new synergies between material assembly, spatial organization, and the search for new human environments, through "motile, mutable and feedback-based relations between habitat and inhabitant that yield diverse and intense social interactions" (*VV*, 7). Such performance-oriented design based on the biological paradigm continues to mechanise the organism as a model for computational processes, and, as such, clearly remains an instrumentalist approach to design. Hensel and Menges themselves admit this: "morpho-ecological design concerns an instrumental approach, making form and function less of a dualism and more of a synergy that aspires to integral design solutions and an alternative model for sustainability" (*VV*, 7).

A frequently unarticulated desire in a performance-oriented organismic model of the digital design process is to reconfigure the status of the human creative designer within that process. Theorists, researchers and practitioners represented by Hensel and Menges attempt to place some distance between the human progenitor and designed forms and processes, by investigating the "creative" forces internal to material and environmental systems. However, an uncomfortable nominalist position emerges here as the models, concepts, abstractions, processes, and machines—what can be called the technological in general—philosophically rely upon a degree of teleology (think what is implicit in the concepts of "performance" and "fit"). Such teleology forces the argument to circle back to "function" or "use" as ontological conditions of forms of being, referring the issues of form and organisation back to a humanist position.

The human-centered or anthropocentric appropriation of the biological paradigm by creative practices, which proceeds from an instrumental model of technology as the static and stable materialization of human thinking, shortchanges the potential contribution of the biological. In fact, invoking the biological paradigm opens the way for biology to overturn the mechanistic model of technology and the physical world. There is a reverse relationship at play, where technology and technological abstractions, methods, processes, and objects within aesthetic practices can be shown to be evolutionary and irreducibly biological. Technological knowledge and objects possess an internal "life-force" immanent to associated systems, whether machinic or epistemological-conceptual, physiological, or libidinal-affectual. My argument here follows a possibility of neither seeing technological knowledge as already given, nor seeing technological machines as concretizations of that knowledge; rather it is to see them as evolving in parallel with entities, bodies (whether human or nonhuman), objects, and other machines adjacent to them. A further argument is that the evolution of technological knowledge and objects toward individuated entities, or, more accurately, toward their effectuations as recognizable technological beings, occurs like biological entities. Design practices that recognize the biological comportment in the genesis of technological objects make contingent the very *how* of knowing itself, which has, in the history of Western thinking, been colored by a mechanistic model. Such practices rework the philosophical connections between the biological-human and the technological-nonhuman worlds.

The discussion that follows further demonstrates that the biological paradigm provides a very different demonstration of how information or knowledge is immanent to the definition of human life. Information and information transfer provide a systemic basis to reconceptualize human agency as a process of emergence between the human and the non-human technological. This process is however far from indeterminate, but is a cumulative process of various local re-combinations of disparate series in symbiotic becomings. Deleuze and Guattari see that re-combinatory processes as driven by the "internal resonance"[7] of an evolutionary system, and that their notion of intensity relies upon a molecularity of the individuating entities, which allows connections between larger phyla of entities without much regard to molar or already-individuated organisation (*DR*, 13).

I proceed in this chapter to draw upon the work of Georges Canguilhem and Gilbert Simondon, two philosophers who have themselves been influential in Deleuze and Guattari's transcendental empiricism and radical materialist philosophy, to identify the biological as an ontogenetic condition within technological objects, processes, and knowledge of contemporary digital practices, which are all immanent to the very ontogeneses these design practices helps to identify.

THE ORGANISMIC EVOLUTION OF THE TECHNOLOGICAL MACHINE

In his famous essay "Machine and Organism," Georges Canguilhem argues to reverse the propensity of reducing living organisms to technological machines. He suggests that the structures and functions of organisms can be used instead as a reference to explain the historicity of ongoing machine constructions. That is, technological concepts, techniques and objects are more than the results of scientific activity; they are in themselves irreducibly biological.[8]

Biological science and technology have a reciprocal relationship whereby attributes of living things are transposed onto machines or objects of human and industrial manufacture. This is in order to explain growth and differentiation of technological systems and their operational tolerances, specificities, and redundancies.[9] Canguilhem writes that, "the construction of machines involving authentically biological notions cannot be understood without revising this view of the relation between science and technology."[10]

The ontological status of technological objects becomes not immediately graspable: the technological, as materiality in constant movement, can be said to be, somewhat, "meteorological." That is, the evolution of technological objects does not have necessarily an immediate and closed connection to the scientific or functional imperatives that purportedly underlie their invention as tools. Rather, the structure and development of technological objects implicate corresponding processes of interaction and exchange with other objects, by virtue of their spatial and temporal adjacencies, or mere passing, as climate—if we continue the metaphor—is emergent from the interactions between complex systems of pressure, humidity, and topography, as each passes another in a constant unity of becoming weather.

A biological conception of technology points to the question of its origins or genesis. Biology itself was grappling with the idea of genesis until the discovery of cell theory in the mid-1800s. At this time, the question morphology—which is as much a question of a technological object as a biological one—plagued biology and the then new disciplines of embryology and physiology: is there an originary form from which all subsequent forms stem? At that time there were two prevalent views: the living mechanism was either associated with a transcendental force behind the form (the creative genius or the machine builder), or "preformation," the notion of an already formed seed within the seed, or a machine within the machine that self produces, and so on *ad infinitum*. When Caspar Wolff, who provided the foundations of modern embryology, showed in his 1759 dissertation that the development of the organism involved a series of non-preformed structures, namely the process of *epigenesis* where organs are formed in differentiated layers from undifferentiated cells, it became necessary to restore responsibility for the organism's organisation to the organism itself. This means that there are formative tendencies within an organism, they develop in a certain way depending on the interactions with the milieu and with themselves. Rather than the mechanistic model of biology, whether relying on a transcendental builder behind the machine as an efficient cause, or the preexisting pre-formed being after which it is modeled, we need to look at the machine not as a finality with finite parts that require a progenitor, but the machine in itself as a desired series of operations fit for and to itself. Machines, with their co-evolving parts, develop specifically in accordance to a definite *idea* that transacts the event of its becoming. This *idea* is not transcendental but incidental, acting as a controlling device for the activation of form that proceeds to operate in a way coordinated by a series of mechanical interconnections.

This idea, which transacts the becoming of organisms, is none other than information. And since the advent of cell theory and then genetics, this information is DNA, which does not provide the form or structure to the organism, but its organic unity. Likewise, a machine is characterized as more than its mechanical and chemical properties; its development occurs under peculiar conditions in accordance with specific epistemologies. The organic cell, while being a basic unit in the living body, is an organism in and of itself, that itself reflects the exigencies of life. However, to Canguilhem, the living machine is made up of these individuated and individuating cells, it does not exist for itself as a superstructure, nor for the other organs and systems, but for the cell itself. The living body becomes the collective means by which cells can express their individuality. Canguilhem therefore sees little difference in the physiology of the machine: it is made up of parts, which depend on the whole, existing solely in order to maintain it (*VR*, 299).

TECHNOLOGICAL BODIES AS PARTIAL CONCRETIZATIONS

Canguilhem's theory of the organism and machine overturns the parts-whole relation. In biological organisms, cells as parts are in a sense more than the whole, and they are also themselves less complicated wholes. As individuating beings, cells in their epigenesis provide the organisational nuance of the organism. Or stated in another way, biological or organismic organization is not the interrelation of cells as parts to a whole, but a totalisation of cells as individuating (and not already individuated) beings. Likewise, machinic parts only operate within teleology on condition that they are primed to do so by virtue of their on-going ever-changing totalisation. Machinic or biological parts are no longer precursors to the success of an organism or entity, but rather the appearance or visibility of the biological and the technological that does not stem from the totalization of parts-whole behaviors.

When we look at cells as organic parts, and machines as technological parts, we are not seeing the origins of the organism, but the organism as a snapshot or a momentary framing of a passing process of individuation. This is not to say that substantive structure and form are completely negated as the basis of organismic composition, but that, at a particular moment the substance of composition is always already transacted by forces, namely information in technology, and genetics in living structures. Conceptual or hereditary information, transferred through substance helps organize the substance at that particular moment in a genealogical chain, the ends of which we cannot see. That is, we cannot see the birth of organisms but only their continuation: the literal births of organisms are but a part of the infinite process of variation and becoming of parts.

Like precipitation in weather, whose duration and intensity is the shape of what has come to pass, or what has passed through, there are certain forms that gel, and information that transact, in the development of technological objects and processes which affect the very way in which technological objects morphologically appear in the world. This occurs when a particular part of a machine becomes functionally over determined, for example fins in machines that originally acted as cooling foils that subsequently became structural; or certain features persists in contemporary objects even if they have an obsolete functionality, for example the "dial" tone in digital telephones.

French philosopher of technology Gilbert Simondon calls this process "concretisation" in technological development;[11] and concretisation is a process that is always only ever partial or incomplete. We can imagine the technological milieu as a multitude of particulates in the atmosphere, which variously recombine and separate by their adjacencies and transactions, some concretizing to become visible, as in rain or snow, others remaining as amorphous as haze, and others again, like ozone, whose presence is felt in their absence. All transmit information on place, from one place to another, by virtue of their trans-formations as assemblages. In the technological world, these elemental particulates are the machinic objects, which in their emergence, transduct information—the very information that composes, or informs, the birth of technological individuals. That is, the individuality of technological objects lies in the functional diagram they sketch out for themselves as they emerge in a group or series of corresponding material concretisations.

To put this argument about technology within the framework of contemporary biology, Canguilhem sees technological relations, or the connectivities between technological objects, as organismic. That is, "within an organism, there are no distances: the whole is immediately present to all the (pseudo-)parts" (*VR*, 318). And the organisation of the parts that is the organism occurs for the reception and transmission of information. Contemporary biology, as a science of heredity through the systems and behavior of the molecular structure of DNA, requires us to understand that what constitutes life in living things, cannot proceed through Euclidean space, but requires a science of order and combination, or topology. This science of life no longer resembles an architecture given by habitable geometric space, but a grammatical or syntactic space that needs to be decoded through the process of transmission before it can exist as space for living.

It is this science of ordering that provides the visibility of how information—in short, genetic code—is produced, transmitted and received through form and structure, which in turn is modified through the process of transmission. To see it another way, organisms as an assemblage of parts, or organs, are concretisations immanent to information transfer; and that the structure and form of the organism, like the structure and form of technological objects, are momentary or partial solidifications or manifestations of information transfer. The organism therefore "evolves" to receive and transmit certain kinds of information; and the information that the organism is not structured to receive might as well not exist as far as the organism is concerned.

What I want to emphasise in this essay is that the transduction of information occurs across adjacent but a-parallel processes of individuation. When this point is reflected upon in context of contemporary digital design, whose processes of individuation hover ambivalently, between human authors and technological automata, between human-crafted objects and computationally generated ones, my concern then surrounds human citizenship in the techno-design milieu when the biological paradigm is invoked. The task here is to place the process of human individuation alongside that of technological objects and processes, and to use biology to explain the transduction of information and affect from the human to the machinic.

INFORMATION AND INDIVIDUATION

The efficacy of a critique of molecularity in digital design lies in the status and location of information within the process of individuation. To Mark Hansen, it is location of information that radically differentiates Deleuze's "internal resonance" from that which he borrows from Simondon. In Deleuze's "internal resonance," pre-individuated entities within the system, whether they are forms, objects, bodies, concepts, or information, remain always partial and always becoming. The entities are organised by the fact of their potentiality, akin to how potential sits within a thermodynamically entropic viral system (apologies for mixing physical and biological metaphors here). Deleuzean virology therefore dissolves everything into a molecular level, making up a realm of ontologically partial but consistent particles in constant individuation. Such virology places no regard for the possibility of pre-established molecular organisation because the consistency of all particles as epiphenomenal and autonomous is given by the fact that they are all attributes of the one intensity (and here Deleuze is influenced by the Spinozist One-being that is transcendental). In short, what conditions actualisations within a Deleuzean system of individuation is a *transcendental* intensity, which is effectively an informative force, located *outside* of the system.

However, what we can see from a closer look at Simondon's process of individuation is a foregrounding of the *form* of the individuating body or entity in the empirical plane, as central to a systematic ontogenetic process.

Deleuze uses (pre) individuation to maintain the intensity of being which then is determined (through the process of individuation) in extensive form. Simondon, contrary to Deleuze, introduces an intensity which simply cannot be separated from the specific processes informing the emergence at the concrete empirical field, which is where complexity emerges, including the processes of morphogenesis that govern the creation of organisms like human beings. Form and individuation are parallel and co-evolutive. For Simondon, individuation cannot be reduced to something that pre-exists (for example an *a priori* formal type, or systemic intensity) this *same* individuation. There is no ontologically unified substance, form or matter outside individuation. To say it in another way, the process of individuation is primordial to any substance that undergoes individuation, so we "grasp the entire unfolding of ontogenesis in all its variety, and to understand the individual from the perspective of the process of individuation rather than the process of individuation by means of the individual," that is, without restricting it to the production of individuals as a movement from a primordial to material state (*ZI*, 299–300).

Furthermore, Simondon's process of individuation does not rely on wholeness, but upon a radical instability:

> Individuation may be thought of as temporary resolutions taking place in the heart of a *metastable* system rich in potential. The system harbors a certain incompatibility with itself, an incompatibility due to the impossibility of interaction between incommensurable terms of extremely disparate dimensions. [12]

Simondon's theory highlights the empirical determinism in the genesis of the biological or technological entity. The central mechanism for compossibility between different bodies or entities is, for Simondon, a flow of information between one body or entity and the next. This is the concept of *transduction*, that I have already mentioned, which for Simondon denotes a physical, biological, mental or social informational process, propagating within a given area, through a structure.

Each structure then serves to constitute the next one. And at the very time this structuration is effected, in tandem with it, a progressive modification is taking place. Transduction therefore furnishes a principle unity as "nonidentity of the being with itself" (*ZI*, 312): an otherness immanent to self-ness. Its dynamism derives from the metastability of the system moving out of step with itself at each encounter and developing further dimensions upon which it bases its structure. Hence, the biological or technological entity—whether the human designer, the computer software, the phylogenetic forms, or the fabrication machines—is not reduced to a pre-individual singularity, but it is a "living" being—a life—which acts as "a node of information, transmitted inside itself" providing the individuating entity with its power of ontogenesis.[13]

Simondon's concept of information provides the basis for describing how information works in the material plane to connect scientific ideas, and technological knowledges into adequate ideas that increase the potential of the body or entity to act; that is, it brings information into the ontology of the human individual and the technological machine. In the context of the biological paradigm in a technological field such as digital design, biological knowledge is therefore transductive in that it is transmitted within itself to elicit change.

The biological does not provide the thing that exists prior to individuation, whether it is the animal, the human body, or the technological object, as individuation cannot be reduced to something that preexists individuation since that would restrict analysis to already individuated beings and obscure the process of ontogenesis. Biology works as both a logical or metaphysical notion applied to identify the process of ontogenesis, and is itself ontogenetic.

The transduction of information, as demonstrated in Simondon's process of individuation, is thus the conceptual baggage that comes with the appropriation of the biological paradigm in the field of digital design. Genesis—whether it is the development of differentiated entities from undifferentiated structures that may not resemble them but contain the folded sequence of implicate codes or diagrams for the effectuation of those entities depending on the interactions with the milieu and with themselves (epigenesis), or the development of differentiated entities from *within* others in a process involving the transduction of information from one entity to another, subscribing to a development of form through contagion (endosymbiosis)—is no longer unified by the transcendental existence of potentiality, which for Deleuze and Guattari is the "internal resonance" of an entity or being. Such intensive force is given by the molecularity of continuously transforming entities in a plane of existence that is *a priori* and therefore outside of the process of individuation, and always already available for effectuation. It is therefore more accurate to say that the individuated or differentiated form, whether human, nonhuman or machinic, of biological and technological beings, is neither local nor global, neither molar nor molecular, because all of these need to be theorized as ontogenetic: continuously being brought into existence through practices that change the conditions.

CONCLUSION: WHAT IS BIO-TECHNOLOGICAL DESIGN?

What we have today in contemporary digital design is a complex conjunction between the biological and the technological. By drawing on phylogenetic evolution, topological equivalences, and emergent properties in form generation and spatial thinking, design technology believes itself to belong to a continuum with the biological by demonstrating a "vitalism," or expressing a "life-imperative," so to speak.[14] Concepts such as "atmosphere," "virus," "cells," and "evolution" appear with some frequency as descriptions and justification of processes and, as such, are relegated as natural-biological almost without qualification. Equally, technologies associated with the production of algorithms and the writing of code in design, mathematical modeling of responsive structural and material systems, or graphical and imagistic visualisations of environmental or demographic flux, are seen as possessing direct analogical relations to natural processes. So, on one hand the biological paradigm identifies ontogenesis within design—its conditions of (re)production and development—as technological; and on the other, computation and associated discourses articulate technological objects and processes of architecture and design as inherently biological and possessing biological behaviors.

The argument here follows the possibility of not seeing technological knowledge as already given, nor technological machines as concretisations of that knowledge, but as evolving in parallel with bodies (whether human or nonhuman), objects and other machines adjacent to them. A further argument is that the evolution of technological knowledge and objects toward individuated entities, or more accurately toward their effectuations as recognizable technological beings, occurs like biological entities. Unless we take seriously the implications of the biological world appropriated by design technology, we shortchange the potential contribution of the biological paradigm to open the way for an overturning of the purely mechanistic model of technology and the physical world.

The appropriation of the biological in new technologies within creative practices draws heavily on the work of Deleuze and Guattari and their use of biology and physics in order to analyse, intervene, if not celebrate, the self-ordering micropolitics of contemporary systems, whether biological or technological or global. By rehearsing the work of two philosophers, namely Canguilhem and Simondon who Deleuze and Guattari use to construct their radical materialist philosophy, I hope to have shown that the comportment to nature and the natural as molecular, in the Deleuzo-Guattarian project causes its appropriations in creative practices to have a tendency to homogenize, if not depoliticize, notions of both difference and continuity between the biological and the technological. That is, the tendency is for such biophilosophy, when appropriated in creative practices, to paradoxically reify the very smoothness or molecularity of the systems in which it attempts to intervene. In this context, to see technology as a dynamic concept that grasps the biological with its partial concresences, and demonstrating the essential role that molar rather than molecular consistency plays in genesis, is an attempt to reclaim an ethico-aesthetic role for design practices.[15]

NOTES

1. See Gilles Deleuze and Félix Guattari, *A Thousand Plateaus: Capitalism and Schizophrenia*, trans. Brian Massumi (Minneapolis/London: University of Minnesota Press, 1987), 164. Hereafter cited as *TPC*.

2. See Manual DeLanda, "Nonorganic Life," in *Zone 6: Incorporations*, ed. Jonathan Crary and Sanford Kwinter (New York: Zone Books, 1992). Hereafter cited as *ZI*.

3. *TPC*, 511. Deleuze and Guattari's quote reads in full: "Abstract machines consist of *unformed matters and nonformal functions*. Every abstract machine is a consolidated aggregate of matters-functions *(phylum and diagram)*. This is evident on a technological 'plane': such a plane is not made up simply of formed substances (aluminum, plastic, electric wire, etc.) or organizing forms (programme, prototypes, etc.), but of a composite of unformed matters exhibiting only degrees of intensity (resistance, conductivity, heating, stretching, speed or delay, induction, transduction. . . .) and diagrammatic functions exhibiting only differential equations or, more generally, 'tensors.'"

4. *TPC*, 69. Deleuze and Guattari continue: "If one begins by considering the strata in themselves, it cannot be said that one is less organized than another. This even applies to a stratum serving as a substratum: there is no fixed order, and one stratum can serve directly as a substratum for another without the intermediaries one would expect there to be from the standpoint of stages and degrees."

5. Michael Hensel and Achim Menges, *Versatility and Vicissitude*, *AD* (March/April 2008), 11. Hereafter cited as *VV*.

6. Michael Hensel, "(Synthetic) Life Architectures: Ramifications and Potentials of a Literal Biological Paradigm for Architectural Design," in *Techniques and Technologies in Morphogenetic Design*, *AD*, eds. Michael Hansel et al. (March/April 2006): 19–20.

7. Internal resonance is a term Deleuze appropriated from French philosopher Gilbert Simondon. See Gilles Deleuze, *Difference and Repetition*, trans. Paul Patton (New York: Columbia University Press, 1994), 246, 318n25, hereafter cited as *DR*. See also Mark Hansen, "Internal Resonance, or Three Steps towards a Non-Viral Becoming," in *Culture Machine* 3 (2001), accessed November 22, 2009, http://www.culturemachine.net/index.php/cm/article/viewArticle/429/446.

8. Canguilhem considers "technology as a universal biological phenomenon and no longer simply an intellectual operation to be carried out by man" (*ZI*, 63).

9. See William Taylor, "Building on Transience: Tolerance and the Subjective Dimension of Technology," in *Techniques and Technologies, Transfers and Transformation,Association of Australasian Schools of Architecture (AASA) International Education Conference* (Sydney: University of Technology Sydney, 2007). Hereafter cited as BT.

10. Georges Canguilhem, *The Vital Rationalist*, ed. François Delaporte (New York: Zone Books, 2000), 297. Hereafter cited VR.

11. See *ZI*, 296–319.

12. *ZI*, 300. Why then does reality as we know it seem stable and not metastable? The stabilities within this reality are in fact only relative as the difference between stratifications and fluid raw matter-energy is a question of the speed and slowness of transformative becoming (morphogenesis) within a single physical system. That is, the system appears stable only because the speed of transformation is low, and in fact there is no absolute distinction between form and matter. Deleuze and Guattari call this the process of stratification. See *TPC*, 40. See also *ZI*, 143. Deleuze develops the notion of slowness and speed from Spinoza's movement and rest immanent to all individuals. See Deleuze, *Spinoza: Practical Philosophy*, trans. Robert Hurley (San Francisco: City Lights Books, 1988), 123.

13. *ZI*, 306. Simondon's full quote reads, "The living being can be considered to be a node of information that is being transmitted inside itself—it is a system within a system, containing *within itself* a mediation between two different orders of magnitude."

14. A further argument, which can be made, one that I have made elsewhere, is that the rationale for biologically inspired creative practices alludes to the view that human experience, expressive of the human individual as a biological phenomenon, is somehow fundamental to aesthetic practice, thus implicating aesthetics itself as biological phenomenon. Human experience here becomes conceived not anthropologically but generically. That is, experience does not pertain specifically to the human individual, but to a general "natural" phenomenon. In the equivocation of experience and nature within the biological paradigm, aesthetic thinking and practice can hence be opened up to natural processes, and its affiliated scientific conceptions, including for example, the application of Darwinian adaptive evolution to the development of aesthetic comportment. See Stephen Loo, "Responding by Mimicry, or Three Frames towards Becoming a Visual Animal," in *Visual Animal: Crossovers, Evolution and New Aesthetics*, ed. Ian North (Adelaide: Contemporary Art Centre of South Australia, 2007).

15. A developed version of this chapter is published in Loo, Stephen. "Emergent Molarities: Resistances on the Molecular Plane of Biology and Digital Architecture." *Architectural Theory Review* 17, no. 1 (2012): 60-75. Copyright permission granted by Taylor and Francis to reproduce parts of this work.

Chapter Three

Design, *Second Life*, and the Hyper Real

Lisa Dethridge

Over the last century we have observed yesterday's science fiction become today's virtual reality. Traditional technologies—phone, cinema, and TV—have converged within the global interconnectivity of the Internet. Human biology and intelligence have converged with artificial intelligence and robotics. In the arts, entertainment, medicine, the military, science, and entertainment we now project multiple versions of the self onto an array of traditional and new virtual environments. This means we now coexist in both real and virtual spaces—onscreen and online—with a variety of artificial forms of life and intelligence. This chapter explores the design epistemology of virtual worlds via the case study of *Second Life*, an interactive, 3D graphic media platform. *Second Life* is a latest-generation multi-user virtual environments (MUVEs). It is available globally to Internet users who meet online to socialize, to create, and to share digital identities, objects, and locations.

What is the territory that is represented or simulated by *Second Life*? How do we understand the forms of virtual "life" that are available to designers for research within this enriched media platform? How do we define *global* and *local* within this cyber-context? To explore these questions we will define virtual worlds and apply Jean Baudrillard's theory of the simulated image to explore aspects of design epistemology in *Second Life*. We will employ Baudrillard's notions of *hyper reality* and *simulation* to address virtual worlds and other artificial life forms such as robots and artificial intelligence programs. We will observe how the pseudo-photographic 3D graphic images in *Second Life* may be married to artificial intelligence and robotics programmes to achieve an uncanny and deceptive fusion of the model and reality.

So how are we to understand and define virtual worlds? Virtual reality is defined as that which is not real but which may display qualities of the real.[1] As Margaret Wertheim points out, virtual cyberspace may feel like a new concept but its conception has been framed by thousands of years of spiritual and scientific thought. The concept of the virtual world may be compared to historical models such as Plato's spheres and medieval notions of heaven and hell.[2] Seen in this light, virtual worlds like *Second Life* are a kind of synthetic or cybernetic universe existing parallel to the "real" world (www.secondlife.com). We call such worlds "persistent" as they are not switched-on or off like videogames or TV.

Multi-user environments evolved from computer games but have gone beyond a rules-based, goal-driven, win/lose game scenario. They exist on the web as social spaces and are designed for social and economic networking. They continue online in real-time, twenty-four-hours a day. By definition then, reality can be posited here as a floating term inside virtual environments.

Virtual worlds are becoming places to conduct serious business, to invest in brands, interact with products, and to simulate systems before deploying them in the real world.[3] Virtual worlds are also a focus for medical practitioners, biologists, and neuropsychologists who use interactive technologies for therapy, training, rehabilitation, and research.[4]

The evolution of virtual media is most evident in arts and entertainment. Just as the twentieth century formed global audiences for screen culture, the twenty-first century offers an addictive labyrinth of pleasures available in a range of 3D graphic formats and interfaces including Xbox, Wii, and multiplayer online games such as *World of Warcraft* and *Penguin Hotel*, referred to as virtual worlds or MUVEs.

This chapter will illustrate how the global, online culture of *Second Life* is a working prototype of virtual worlds, which media theorists suggest will become common work and leisure places of the future.[5] One problem in the field concerns epistemology. As designers perfect the realism of the 3D graphic interface, there is an increased convergence between biology and machine or robotic forms, or between human and artificial intelligence. The psychological, ethical, and technical question of who animates whom, in the virtual world, is becoming more pressing.

Using French philosopher Jean Baudrillard as a guide, the discussion now examines *Second Life* to question notions of truth, location, identity, and authenticity as they operate inside these virtual environments. It is clear from the following research that aspects of self are extended in this realm, and that forms of global and local knowledge may become conflated or even interchangeable.

METHOD

We will outline several conditions that govern the *Second Life* virtual world and look closely at the relationship between the human user and their online avatar. Jean Baudrillard's theories of "simulation" and of the "hyperreal" provide insight into our perception of virtual worlds. Baudrillard argued in the 1980s that new media technologies change our perception of reality. His premise is based on the historical and technical difference between media production methods that represent reality and those that simulate reality. Baudrillard recognised that computer-driven media simulations generate a complex layer of illusion. In *Simulations* he asserts that media simulations like *Second Life* may confuse our ability to distinguish, both visually and conceptually, between real and fantasy elements. In *Simulations*[6] and *Screened-Out,*[7] Baudrillard suggests that in conditions of computer-driven simulation, the problem of "truth" is lifted out of our hands and remains floating within the computer codes of the hyperreal zone (*S*, 97).

As with his contemporaries Michel Foucault, Gilles Deleuze, and Félix Guattari, Baudrillard is concerned with epistemology, or the science of how we know what we know.[8] Baudrillard surveys the history of European art and the image, focusing on various pictorial modes in the experience and expansion of visual space. He considers the shift from the flat 2D imagery of ancient art to the illusionistic penetration of the 3D picture plane investigated by Renaissance artists. In late twentieth-century industrial cultures, Baudrillard observes that simulation as a condition begins to breed dangerous confusion around computer-generated images and situations. This media condition results from "the seductive power of endless stimuli" (*S*, 139).

Another key term relevant here is that of "convergence," which relates to current shifts in the configuration of media technology. There are five aspects of convergence that provide a context for this discussion of design in virtual media:

1. Technological convergence of digital media;
2. Disciplinary convergence of arts and sciences;
3. Social convergence of individuals and groups using P2P, online networks;
4. Metaphorical convergence of global and local cultures;
5. Psychological convergence of human/avatar.

We will discuss these aspects of convergence with a focus on the relationship between human users and their online avatars, the graphic representations of self that appear onscreen. We will then speculate on how rapid changes in technology may facilitate larger shifts in epistemology, that is, in how we "know" what we know in virtual worlds.

The central question is: What notions of reality are represented or simulated by *Second Life*? How do we understand the forms of virtual "life" and "territory" that are available to designers for research within this enriched media platform? How do we define *global* and *local* within this context? This inquiry may be relevant not only to scholars but also to the producers, programmers, and creators of virtual and cyber environments across a wide range of fields. Such study is important in order to assess the future of a society that may be destined for a variety of artificial intelligence forms, including robots, which have already appeared in *Second Life*. It is therefore apt to observe this case study as it represents a test-bed for future virtual cultures.

DEFINITION OF *SECOND LIFE*

Second Life can be defined as a multi-layered information platform—or meta-platform—for a range of applications including 3D graphic, audiovisual and communication tools. Databases and high-powered computer servers drive all on a matrix of networks that are distributed globally.[9] *Second Life* is not a game with rules and objectives, but a social environment that incorporates a sense of play. *Second Life* provides users or "residents" with shared, online spaces that mimic or simulate those of both real life and of fantasy. It takes the visual form of a geographical "world" complete with islands, buildings, landscapes, universities, corporations, shopping malls, and social venues.

Currently, users log on from all over Europe, Scandinavia, North America, Australia and parts of Asia, nations where hardware and broadband penetration are not an issue. They "immerse" themselves in, or "inhabit" the synthetic, 3D graphic space, interacting with other avatars and engaging with applications including text, photography, movies, interactive animations, 3D sculpting and chat. This allows for a wide range of personal and collaborative activity (*SLO*, 8).

Each *Second Life* user is represented online by their colorful graphic avatar, the digital graphic image of "self," which is part of the graphic user interface. The avatar allows a user to experience vicarious vision, public visibility, physical motion and interactivity within the virtual world. The term avatar comes from the Sanskrit for a divine being who has descended to earth.[10] As we shall see, this all suggests that the avatar acts as both a psychological and physical link between the real and the virtual worlds. To understand the complexity of programs and output that are produced by this system we may briefly take account of the larger historical, social, and technological factors that are driving it.

TECHNOLOGY AND CONSCIOUSNESS

Baudrillard's vision resonates with Marshall MacLuhan's famous maxims: firstly, the medium is the message, and secondly, electronic media technologies are like extensions of our nervous system allowing us to project our consciousness into the world.[11] How might technology act as an object or screen for psychosocial projection? Baudrillard wrote *Simulations* in the same year as science fiction novelist William Gibson was winning awards for his novel *Neuromancer.*[12] Both authors draw attention to the psychosocial dimension of simulation. Gibson coined the term "cyberspace" to describe the visible surface or interface of a virtual world. This computer-generated illusion is a kind of mass, consensual hallucination, perceived as "real" by the users who live vicariously through their avatars. Back in 1984, both Gibson and Baudrillard were concerned with an almost psychotic convergence between user and technology within virtual worlds.

From an historical perspective, the designer may be reminded of the archetypal human urge to simulate and model our physical reality. The fact that we do so while peering into dark space is not new. Artists and philosophers, magicians, shamans, hunters, and astrologers have always projected human consciousness into mythical heavens using images and tools involving light and shadow. They did so to build myths about avatars, gods, and spiritual and extraterrestrial visitors from other dimensions who visit this one and offer knowledge.[13] It is clear that humans have always projected symbolic representations of self and objects into some abstract context. Certainly we imagine the future in order to get there. This ability may be one of the survival and problem-solving strategies that is common to all cultures and epochs.

In this context, the avatars of *Second Life* and other 3D virtual worlds are genealogically linked to the shadow puppets of prehistoric cave dwellers. An anthropologist may compare the avatar to the totems, dolls, tribal fetishes, automatons, robots, and icons of traditional and modern societies. Humans project their consciousness onto such inanimate objects, which are symbols of "magical thinking" and represent aspects of the "other worldly" for cultures across the globe.[14] The designer of virtual worlds may then be aware that the avatar allows us to project ourselves into another visual graphic space, which represents another symbolic or imaginary realm.

Susan Sontag points to a fundamental faith we have developed in the twentieth century: that a photograph is an authentic depiction of reality by virtue of its documentary nature.[15] In twenty-first century virtual environments, the imagery is not photographic, but computer-generated. However, within 3D virtual worlds like *Second Life*, the conventions of photography concerning light, shade, and composition suggest, on an almost subconscious level, that we are looking at a "real" photograph of a "real" object, person, or event. Baudrillard would ask how are we to distinguish authentic photographic documents and images from artificially generated ones?

The problem raised by Baudrillard's theory of the sign is that in creating more realistic, or in his terms, hyperrealistic, models to plan our future, we may lose our capacity to distinguish between imaginary models of reality and the reality itself. Pseudo-photographic 3D graphic images achieve an uncanny and deceptive fusion of the model and reality. This in turn may lead to an epistemological confusion, which is the basis of Baudrillard's theory. We see and somehow believe the evidence of our senses despite the fact that they refer us to an unreal, or in Baudrillard's terms hyperreal zone.

This discussion of epistemology forms a background for the central questions: What notions of reality are represented or simulated by *Second Life*? And, how do we understand the forms of virtual life and territory that are available to designers for research within this enriched media platform? As noted above, there are five key aspects of "convergence" between technology and human psychology, which provide a context for this discussion of design for virtual media.

1. Technological Convergence

On the one hand, convergence relates to the physical linkages of various technologies and applications within larger and larger systems such as telephone, television and the cinema. It is beyond the scope of this paper to detail this well-documented convergence beyond observing the increased aggregation of media content across net-based meta- platforms like *Second Life*, which in turn host or link to a range of other secondary applications and tools.[16]

2. Disciplinary Convergence of Arts and Sciences

Convergence also relates to the increased crossover and sharing of 3D graphic tools in different sectors of the arts and sciences. We see this in news, entertainment, the military, forensics, law, education, and science. 3D virtual technologies allow for display imagery of unprecedented audiovisual realism. When such technology is used across scientific and legal contexts we may need to focus clearly on the location of truth and authenticity within such media representations.[17]

3. Social Convergence of Individuals and Groups using P2P Online Networks

Individuals and groups now log onto virtual screen interfaces at the office, the school or university, the factory, the design studio, or in the living room. We chat to friends on Facebook; we share images on Flickr or YouTube and capture elements of an entire audio/visual lifestyle on MySpace. We also collaborate and share work files on the corporate intranet. Digital convergence has become a crucial pre-condition for social networking.[18]

These networking sites exemplify the postmodern “text” whereby aspects of an entire story-world are created not by a single author or corporation, but by users who often collaborate to generate their own content (*VR*, 318). This process is enabled by digital files, which are easily reproduced, repurposed, distributed, and presented. As Lev Manovich reminds us, new digital artforms are essentially the manipulation of variables within a database.[19]

The virtual 3D world offers a publishing tool where artists and creators distribute virtual goods or information. *Second Life* residents and visitors export digital information, movies, images and objects into the “real world.” They create a personal archive or database that acts as an enriched digital profile to be shared with others both “in-world” and outside. Local *Second Life* culture—digital objects and media produced by SL residents “in-world”—then becomes global as users extend their cultural reach from *Second Life* into the real world via web links.

To facilitate these interactions, *Second >Life* provides an in-world economy where users can exchange real money for “Linden dollars.” Clearly, *Second Life* represents a prototype of virtual economies exchanging and circulating goods and services between real and virtual environments. A growing body of research now considers the impact of this kind of exchange on various legislative and commercial real-world environments and territories.[20]

4. Metaphorical Convergence of Global and Local Cultures

In *Second Life* definitions of global and local are conflated and merged. *Second Life* is based on a metaphor of the globe itself, which mimics the geographical scale and diversity of the planet. The virtual world provides thousands of highly personal or localized audiovisual environments or islands that are linked within an over-arching computer matrix. Various island regions host a range of human cultures and activities that are at the fingertip control of the user. It is a model of global consumer paradise.

Keeping pace with the appetite for what Baudrillard calls “endless stimuli,” the avatar can travel at will within this virtual world, which reflects and distorts the real with amusing fascinations. One can visit the Tower of London, the Sydney Harbour Bridge, gothic castles, seaside gardens, or go scuba diving or sailing. One can design aspects of the avatar’s appearance and engage in countless other simulations of human behaviour and culture in a digital context.

Most of the *Second Life* environment is administered not as a game nor as a business, but as a way of life. *Second Life* residents have town meetings to discuss administration rules; they may attend a seminar at the virtual campus of RMIT University or Harvard; or watch a rock concert sponsored by the British Broadcasting Corporation (BBC). One can, frankly, spend all day in *Second Life* and never be bored were it not for the cramped conditions of the human viewing environment, conditions that will soon change as display technology moves into the next generation.

Our definitions of global and local become distorted in this context of simulated reality. In *Second Life*, people from widely disparate geographical realms meet and collaborate in a shared space of virtual dimensions. In addition, the program provides a range of maps and navigational aids that allow a user to shift positions at will between macro and micro depictions of the world using maps and camera techniques that show highly specific, local detail. *Second Life* offers a seemingly enhanced point of view that is not fixed. The user can observe the avatar's global position in the world with a variety of maps and search aids.

The avatar's vision is depicted onscreen as the output of a camera, which is not tied to the avatar's position on screen, but can zoom, pan, crane and dive with all the alacrity of a high-powered movie camera on a huge gyroscope. This extended power of sight and navigation far exceeds that of the fixed, swivel-vision of the eyeball and retina. The user gains a computer-enhanced vision, which hints at superhuman powers where global scope and local detail are conflated in the one gaze.

Other global aspects of *Second Life* concern the nature of creative collaboration. Objects in this world may be classed as "global" as their creators are geographically dispersed around the world. However, their shared virtual space, where they come together online, is local. Their products then may be classed as a form of local knowledge peculiar to the *Second Life* culture.

This Borgesian labyrinth is like a vast global map seen through a virtual prism. While *Second Life* runs parallel to our own world, distance and time are immaterial in the virtual space that obeys different laws of physics. A virtual sun and moon both rise and set several times each day in *Second Life*. However the hyperreal dimensions seem manageable due to the variety of tools provided for their exploration.

In this way, we may consider the hyperreal as also the hyper networked, as the number and sophistication of tools at my disposal in *Second Life* allows a range of contacts with a huge distributed network. I can fly or teleport myself from virtual locations in Sydney, London, Amsterdam, Paris, or the walled city of Kowloon. I can send instant messages to my friends online or bring them to me instantly via teleport to my location from wherever they are. As a result, the global virtual space of *Second Life* is as easy to negotiate as my local neighborhood.

5. Psychological Convergence of Human/Avatar

As real and virtual worlds converge within the *Second Life* simulation, we observe conditions of "mixed reality." In these new environments, physical and digital characters and objects co-exist and interact in real time. How might designers understand the psychology of the person who is using virtual technology in conditions of mixed reality? Lev Manovich emphasises that under conditions of immersion in virtual screen media, the real world fades away, "you are hardly aware of your physical surroundings" (*NMD*, 79).

Within this context, we observe a form of convergence between our individual selves and the virtual, 3D graphic self or avatar that represents us online. In *Second Life*, the avatar is usually a hyperrealistic image, which may or may not represent true aspects of self. As a user, "I am my avatar." The user selects from animations that simulate body functions—including dancing, sport, and flight. Other applications allow exchange, financial transactions and extensive communication with other users. In this way, online residents of *Second Life* can behave and act out. They build their own virtual houses and design their own virtual lifestyle. They decorate their virtual living rooms with smart 3D graphic furniture and artworks or pay another avatar to walk their virtual dog.

The future challenge may be for designers to form an epistemology that can accommodate the ambiguities inherent in this media landscape. This leads us to further exploration of the forms of character and psychology that are available as design elements in virtual worlds.

ETHICS AND PSYCHOLOGY OF THE AVATAR

In 2008, the BBC reported proceedings of the American Association for the Advancement of Science meeting in Boston. Robotics, Nano-technology, artificial intelligence and virtual reality systems were cited by the U.S. Academy of Engineering as among the most pressing challenges facing humanity in the twenty first century.[21] Each of these technologies has a vital role to play in virtual worlds. Robots—known as "bots"—and other characters driven by Artificial Intelligence (A.I.s), are already familiar fixtures in corporate customer service centers, chatrooms, and of course computer games that feature non-player characters (NPCs) or game-generated characters as an essential part of the play.

The *Second Life* producer, Gary Hayes, is a pioneering designer and architect of virtual worlds. He suggests the relationships we forge now with our avatars represent another stage in the evolution of humans toward "the ultimate in personalization, a digital you." Some of the ethical and psychological constraints of the relationship between you and "digital you" are being tested in *Second Life* with surprising results.[22]

Hayes refers to blogs and forums around the virtual worlds where *Second Life* users discuss the presence of bots, which may be used as corporate spies to gather intelligence on user behavior. These automated characters and avatars are driven by databases rather than by humans and cause social unrest within the virtual space. Hayes calls these bots "invaders" because they inhabit human spaces in the virtual world, but are often not clearly labelled as A.I. by their programmers. He observes this stealthy invasion of A.I. forms is causing "backlash and resistance" among human users. "It has become very difficult to tell now if the avatars have a human or an sql (database) driving them and this is irritating many 'human' inhabitants" (*PMB*, 1).

Hayes reminds us that *Second Life* is a socialized space and not a game. The ambiguity around identity in relation to humans, artificial intelligence and artificial life forms signals a new epoch where robots and humans share virtual spaces and eventually, real space, which may lead to a form of digital mediascape known as "mixed reality." Currently, in conditions of mixed reality we see increased confusion between the real and the hyperreal. In *Second Life* our virtual selves are building a hyper-networked space where "humanoids" and bots are also present. The question then of how bots are accommodated in *Second Life* acts as a test case for issues that may eventually arise in other virtual worlds and in the real world.

It is clear that some confusion in the boundaries between humans, avatars and robots raises ethical issues around digital identity. When is a human not a human? Is an avatar to be treated as a human? Must one always signal one's true human identity when acting as an avatar? Should all bots and A.I.s be clearly labelled as such? Can I inhabit or act within a virtual world as my friend's avatar or is that unethical? Is my digital self liable for the same legal rights and privileges as my real self? How do we protect child avatars? The issue of intellectual and creative property rights is also crucial in this context. Who owns the data around my avatar and around my digital creations?

Research at RMIT University suggests that MUVEs such as *Second Life* contain surveillance devices capable of recording location and chat data by users. Chris Dodds points out that our understanding of the "digital persona" is a model of the person established through the collection and analysis of data relating to the behavior of their avatar online. Administrators can "mine" this data, skimming it for specific content to prosecute in-world offences and monitor the environment's stability.[23]

With the increase of companies establishing a commercial presence in *Second Life*, issues such as personal and corporate privacy, surveillance and espionage are gaining more attention. The collection and distribution of data beyond the *Second Life* environment—such as personal details—is forbidden under the terms of user agreement. However, the proprietors Linden Labs may use aggregated or demographic information from the user base and share it with third parties. This means that while your first life will remain anonymous, information pertaining to your *Second Life* avatar will not.[24]

According to Baudrillard, simulation entails deception when the image is not framed or disclaimed as an artificial reproduction of reality. In his view, forms of simulation are problematic when they are not signalled as such. This may suggest a need for the owners and programmers of virtual systems and robot or bot characters to clearly designate them as programme-generated entities and not human-generated. It may call for an ethics of transparency where humans may eventually need to register or license their avatar's identity.

All of this seems to undercut the very freedoms that users enjoy in virtual worlds and so research is needed to investigate the ethical standards governing social covenants in virtual culture. Virtual worlds like *Second Life* deliver lush, hyperreal, and sensual visions via the computer interface. In future however, that interface or portal into alternate reality may not be clearly flagged.

The movie *The Matrix* by Larry and Andy Wachowski portrays a dystopian future world in which humanity is enslaved by technology run amok. In this scenario we mistake virtual scenarios for "reality" and lose touch with the actual lived reality of our daily lives. In *The Matrix* story world, there is no screen between real and virtual dimensions, but rather an instant injection into the virtual realm via a physical link with the computer. This film is a portend of what may be fast approaching: an epoch in which the screen disappears and we carry the computer inside us, via microchips and mini-circuits, which, like pacemakers and cochlear implants, are designed to enhance our abilities to function.

CONCLUSION

There is clearly a potential for the trajectories traced by the science fiction of the last twenty years to become a reality. Convergence of technologies across platforms like *Second Life* gives designers the ability to create total visual environments. In Baudrillard's terms, virtual worlds allow for aesthetic experience attaining a new hyperrealism, where model and reality seem as one. *Second Life* is a test-bed for virtual worlds, which are an expressive platform for the networked generation. It contains the interfaces already established by a prior century of telecommunications, but facilitates social and economic interactions around new modes of intimacy and mutual responsibility.

There are many benefits for designers working in the *Second Life* virtual world. It is clear that unlike the passive audience of film and TV, there is a higher level of interactivity, creativity and participation for users of the *Second Life* virtual world. Global interactivity allows for global collaboration and exchange, and the convergence of powerful creative tools allows for a range of personal expression, creativity, and publication. There can be positive value in these worlds to extend the potential of fictional and documentary traditional media forms such as books, theater, cinema, and the novel. *Second Life* offers a prototype for designers to explore a range of powerful publishing and media tools. It also represents a prototype virtual economy for virtual goods and services, which can be circulated between real and virtual worlds. It provides designers with opportunities to design avatars, animations, architectures, simulations, and other digital media objects, images, movies and social networks. The convergence of easy navigation-aids allows a designer to shift positions at will between macro and micro depictions of the world using maps and camera techniques that conflate global scope and local detail in the one gaze.

Future research may examine how designers may allow for mixed reality environments where physical and digital characters and objects can co-exist and interact in real time and real space. The larger ethical issues of surveillance, privacy, intellectual, and creative property rights are also a priority. In Baudrillard's terms however, these seductive stimuli continue to raise serious issues for media epistemology in virtual worlds. Research investigating the psychology of play in massive role-play games such as *World of Warcraft* suggests that while technology advances rapidly, our ability to process the epistemological changes may lag behind. Strong relationships can develop in these worlds where people will even marry in avatar form or grieve for deceased friends they have met only virtually. This kind of discussion points to a growing divide emerging between people who think MUVEs are extensions of the real world and those who believe they are not governed by the same moral imperatives.

This discussion demonstrates how the physical conditions for shared symbolic realities are being reconfigured within virtual worlds. It has shown how perception and knowledge of the real can no longer be assumed as one that is shared. Our learned faith in the authenticity of the photographic image allows us to suspend our disbelief in computer-generated imagery. As a result, we become immersed in the experience: What we see is what we believe. When that vision is augmented by sophisticated photo-realistic 3D graphic tools, by robots and other characters and programs driven by artificial intelligence, the location of global and local, of "truth" and reality within the image becomes harder and harder to assess.

Baudrillard's theory of simulation may be useful in the research and administration of virtual worlds that combine social and technological networks in new and powerful ways. His theory reminds us of the need to be vigilant in relation to the logic and ethics we use to manage the real and artificial counterparts of media simulations. Within the hyperreal context of mixed reality, Baudrillard's theory may help us maintain a consistent emphasis on the quality of electronically mediated discourse, on the difference between representation and simulation, and on the clear identification of parties involved in transactions. Baudrillard's theory of the hyperreal may help to critically analyze and clarify the logic of simulated media events, ensuring the maintenance of objective standards of truth and semantic logic.

NOTES

1. Paul Milgram and Fumio Kishino, "A taxonomy of mixed reality visual displays." *IEICE Transactions on Information Systems* E77-D, no. 12, (December 1994). See also William Sherman and Alan Craig, *Understanding Virtual Reality* (London: Morgan Kaufman Publishers, Elsevier Science, 2003).

2. Margaret Wertheim, *The Pearly Gates of Cyberspace* (New York: W. W. Norton & Co., 1995).

3. Mandy Salomon, "Business in Second Life." Paper presented as *Smart Internet Technology CRC* Research Paper, last modified December 22, 2010, accessed June 18, 2008, http://www.smartservicescrc.com.au/Outcomes.html. Hereafter cited as *BS.*

4. Heidi Sveistrup, "Motor Rehabilitation Using Virtual Reality," in the *Journal of NeuroEngineering and Rehabilitation* (December 2004), 1:10. Maureen Holden, *CyberPsychology & Behavior* 8 no. 3 (June 22, 2005): 187–211, doi: 10.1089/cpb.2005.8.187.

5. Lev Manovich, *The Language of New Media* (Cambridge, MA: MIT Press, 2001). Hereafter cited as *LNM.* Jay Bolter and Richard Grusin, *Understanding New Media* (Cambridge, MA: MIT Press, 2000). *New Media, Theories and Practices of Digitextuality*, eds. Anna Everett and John Caldwell (New York: Routledge, 2003). Hereafter cited as *NMD.*

6. Jean Baudrillard, *Simulations* (New York: Semiotext(e), 1983). Hereafter cited as *S.*

7. Jean Baudrillard, *Screened-Out* (New York: Verso, 2002).

8. Gilles Deleuze and Félix Guattari, *Anti-Oedipus-Capitalism and Schizophrenia*, trans. Robert Hurley, Mark Seem and Helen Lane (New York: Viking, 1977). Michel Foucault, *Power/Knowledge: Selected Interviews and Other Writings*, trans. Colin Gordon et al. (New York: Pantheon, 1980).

9. Michael Rymaszewski et al., *Second Life: The Official Guide* (New Jersey: Wiley, 2007). Hereafter cited as *SLO.*

10. Freda Matchett, *Krsna, Lord or Avatara? The Relationship between Krsna and Visnu in the Context of the Avatara Myth as Presented by the Harivamsa, the Visnupurana and the Bhagavatapurana* (New York: Routledge, 2000), 15.

11. Marshall MacLuhan, *Understanding Media* (London: Routledge and Kegan Paul, 1974), 34.

12. William Gibson, *Neuromancer* (New York: Ace, 1984).

13. Claude Lévi-Strauss, *The Savage Mind* (Chicago: University of Chicago, 1968), 254. Hereafter cited as *SM.*

14. Marina Warner, ed., *Eyes, Lies and Illusions* (Melbourne: Australian Centre for the Moving Image, 2006), 8. See also *SM*, 200.

15. Susan Sontag, *On Photography* (New York: Delta, 1977), 18.

16. Darren Sharp, "Digital Lifestyles Monitor," Paper presented as *Smart Internet Technology CRC* Research Paper, last modified December 22, 2010. http://www.smartservicescrc.com.au/Outcomes.html. Hereafter cited as *DLM.* Nicholas Negroponte, *Being Digital* (New York: Vintage, 1995). Hereafter cited as *BD.*

17. Damian Schofield, "Animating and Interacting with Graphical Evidence: Bringing Courtrooms to Life with Virtual Reconstructions," paper presented at *IEEE Conference on Computer Graphics, Imaging and Visualisation*, Bangkok, Thailand, (August 14–16, 2007). See also Maria Boas Hall, *The Scientific Renaissance, 1450–1630* (London: Dover, 1994).

18. See *BS* and *DLM.*

19. Eric Von Hippel, *Democratizing Innovation* (Cambridge, MA: MIT Press, 2005). See also *DLM.*

20. See *BD* and *DLM.*

21. Helen Briggs, "Machines to match man by 2029" *BBC Science Reporter*, BBC News Boston, accessed July 2, 2008, http://news.bbc.co.uk/2/hi/americas/7248875.stm.

22. Gary Hayes, What have you got against my bot? Personalize Media Blog, accessed July 2, 2008, http://www.personalizemedia.com/index.php/2008/02/18/377. Hereafter cited as *PMB.*

23. Chris Dodds, "Avatars and the Invisible Omniscience: The Panoptical Model within Virtual Worlds." Master's dissertation, RMIT University, School of Creative Media (2007): 52, http://www.iconinc.com.au//christo/c.dodds_exegesis.pdf.

24. Joe Rybicki, April 12, 2006 (7:26 PM EST) "The Real and the Semi-Real," 1Up, http://www.1up.com/do/blogEntry?bId=6883235&publicUserId=4553267.

Chapter Fourteen

Hopeful

Biology, Architectural Design, and Philosophy

Chris Smith

The work of Gilles Deleuze is valuable in re-configuring logics to investigate the potentials of subjectivity; of dealing with what Rosi Braidotti refers to as "the living process of the transformation of the self."[1] Deleuze calls into question the interaction between "selves" as occurrences and the forces through which they are actualised. There is, in his work, the privileging of an affective scope of philosophy that is in conflict with the logics of Reason, a valuing of intensity of thought over the jurisprudences of thought. According to Braidotti, for Deleuze thought is "a way of establishing concrete material and semiotic connections among subjects that are conceived in terms of a multiplicity of impersonal forces" (*NSE*, 111). This manner of thinking about subjectivity is also an opening by which interdisciplinarity might be engaged. It is a mode of thinking that involves a valuing of force over form, a concentration not on the analogical equivalence between selves or disciplines, but rather on their intensive and extensive relations. The value of this type of connection has become more important in architecture's recent engagements with both biology and philosophy. This chapter's focus is on one of the margins of that relation: a margin marked by a monster.

The monster will be discussed primarily because such a fictional body signifies an anxious assemblage of diagrams and representations:[2] concrete material and semiotic connections. This chapter is concerned with the configuration of these assemblages and the potential disruptions that such assemblages may cause disciplines. The monster of immediate concern is of the "hopeful" variety. It is a monster that emerges in the work of the biologist Richard Goldschmidt and (re)emerges in the work of Deleuze. Subjectivity, according to Deleuze is a habit and that must be considered not only in terms of occurrence, but also in terms of the forces, which condition its deployment.[3] For Deleuze there is no nucleus of identity but rather dispersed processes configure at the margins of self: the margins, which expose the self to the monstrous.

This chapter focuses on three primary texts: the first is Marco Frascari's *Monsters of Architecture: Anthropomorphism in Architectural Theory* (1991);[4] the second is Goldschmidt's *The Material Basis of Evolution* (1940).[5] I am concerned with the textual construction of the *hopeful monster* in these texts and for the monstrous acts it signifies. Frascari's fetish for the monster relates to what he describes as the "excrescences and orifices" of architectural design and the monstrous subject is deployed against the architecture of humanism. Goldschmidt is interested in the *hopeful monster* as a biological possibility that allows him to attack the Darwinian notion of gradualism in evolution. A third text, Deleuze's *Empiricism and Subjectivity: An Essay on Hume's Theory of Human Nature* (1953), is functional. This text operates "quietly" through the chapter in providing a number of concepts through which the assemblages of the others are explored.

THE MONSTER AND THE "GIVEN" NORMAL MEN

In *Empiricism and Subjectivity*, Deleuze examines Hume's notion of the self. His tactic is one of exploration, but he is more specific in describing his approach to philosophy in terms of an act of unrequited sodomy.[6] Through this forceful act, Deleuze imagines himself and the philosopher upon whom he focuses conceiving a monstrous offspring:

> My way of getting out of it at the time was, I really think, to conceive of the history of philosophy as a kind of buggery or, what comes to the same thing, immaculate conception. I imagined myself getting onto the back of an author, and giving him a child, which would be his and which would be, at the same time, a monster. It is very important that it should be his child, because the author actually had to say everything that I made him say. But it also had to be a monster because it was necessary to go through all kinds of decentrings, slips, break-ins, secret emissions.[7]

The monstrous offspring of the study of subjectivity belongs to both Deleuze and Hume and to neither. It belongs to both to the extent that Deleuze's writing of Hume certainly allows him to articulate a number of the key points of Hume and then to incorporate and advance them himself. It belongs to neither to the extent that in working with Hume, Deleuze is producing that which is of ambiguous origin. It is by way of this ambiguity that Deleuze is able to advance Hume's project of exploration of how the subject is constituted from that which is given.[8] He utilizes the idea of the "given" as an account of how the subject is formed but is not content to solely rely on the given. This is substantiated in multiple passages throughout *Difference and Repetition* (1968), where Deleuze remarks that we must explain the given itself: "Difference is not diversity. Diversity is given, but difference is that through which the given is given as diverse."[9]

Much of the discourse of the monster centres on this construction of difference and the given. The monster disrupts the expectations of similarity and the same; it is a speciation event that disturbs the prior, and importantly it transgresses boundaries. The monster is also valuable in that it makes smooth space of "given" fatherlands and mother tongues. There is a sense by which the *hopeful monster* creates its own sites and its own inscriptions (*TPC*, 12).

In any consideration of biology since the self-confirming conjunction of Darwinism and Mendelian genetics in what is referred to as "the *modern synthesis*" the "given" is the Darwinian thesis of natural selection. According to Goldschmidt:

> The idea expressed in the somewhat unconventional but plastic term "hopeful monster" is not a new one. We may refer back to Darwin, who pointed out that under domestication monstrosities occur which resemble normal structures in widely different animals (*MB*, 391).[10]

In any consideration of the body in architecture the "given" tends to be the classical Vitruvian figure, iconised in Da Vinci's sketch of human proportion superimposed with Euclidean geometry. This dialectical image operates as an emblematic origin for any discussion of the architectural body and as with all origins it remains difficult to escape. There is no lack of will on the part of architects and architectural theorists to depose the Vitruvian "normal man" of architecture. The motivation behind the deposition is based on the figure's inherent link with anthropocentric humanism and mimesis. It is the "normal" body that contemporary theorisations have chosen as the site upon which to inscribe their denunciation rather than the genealogy that the body signifies.[11] The architectural discourse of the body would appear, even in Frascari's text *Monsters of Architecture*, to depart from behind the "given" normal man.

THE *HOPEFUL MONSTER* OF BIOLOGY

1. In Darwin

If *The Origin of Species* (1859) attempts a "domestication" of the monster, then it proceeds by way of two strategies. Both strategies may be regarded as processes of internalisation pertinent to any major science; and neither route is without the significant, if not fundamental, cost of the monster's transgressive potential.

First, the monster is removed from the bounds of Darwin's theory of natural selection by a distinct form of death: the loss of the monster's potential to establish itself as normative. By this strategy, within *The Origin of Species*, the monster is that which can be neither understood as adaptation nor read as an evolutionary motion where, "by a monstrosity I presume is meant some considerable deviation of structure in one part, either injurious to or not useful to the species, and not generally propagated."[12] The result is that the monster is sterilized at the disciplinary limits of the natural sciences. The monster as born termination or specious dead-end is the consideration of the monster as pathological in evolution: If alive, the monster is considered non-reproductive; and if reproductive, the monster would not propagate that which makes it monstrous.

This negation of the monster is a removal of the beast from the bounds of the discipline, and yet, even the above extract from *The Origin of Species* is provocative of a reading of the potential of the monster to be "hopeful." The grammatical clarification that the monstrous is "not *generally* propagated" is to maintain an opening by which the monster could participate in Darwin's theorematic: the second strategy by which Darwin would attempt a domestication of the monster:

> The naturalist includes as one species the several larval stages of the same individual, however much they may differ from each other and from the adult; as he likewise includes the so-called alternate generations . . . which can only in a technical sense be considered as the same individual. He includes monsters; he includes varieties, not solely because they closely resemble the parent-form, but because they are descended from it (*OS*, 424).

The second strategy by which the monster is dealt is one that Deleuze and Guattari might refer to as "appropriation proper."[13] Here, within the theory of gradualism the monster would tentatively make the movement from monstrosity to normative possibility.[14] The discourse of Darwin attempts an appropriation of the monster not because it is *as* the parent but because it is *from* the parent. It is through the evolutionary *gradualism* promoted by *The Origin of Species* that the transgressive potential of the monster is to be dispersed: disseminated thinly across generations to the point where that which is monstrous is to be understood only as accumulated variation. The primary assertions of the "long argument" of Darwin were to be corroborated in a theoretical synthesis with the discourse of Mendelian genetics establishing the genetical theory of selection. The fundamentals of Darwinian theory: that evolution is a dual process where random variation (micromutation) provides the material upon which the driving force of natural selection would push evolution; and that evolutionary change is a gradual process, are significantly substantiated by the synthesis.[15] Stephen Jay Gould suggests that notions contrary to the discourse of gradualism are, under the authoritative self-confirmations of the *modern synthesis*, routinely "corrected."[16]

2. In Goldschmidt

Goldschmidt's text *The Material Basis of Evolution* is received with malevolence by the discourse of neo-Darwinism. The text is widely considered an unfortunate and unnecessary retrogression to *mutationism*.[17] The text suffers, at least in part, for the heresy of Goldschmidt's attempt to free the monster from Darwin's domestication strategies. The destabilizing potential of the monster within the text of Goldschmidt is asserted conscientiously against the apparatuses of the *modern synthesis* in a manner that would prove not merely provocative of potent reaction but generative of the type of ruptures to the major science that incite an enduring and vigilant form of maintenance.[18] The monster re-emerges as "hopeful" in Goldschmidt's discourse of *macromutation* not merely within the framework of the theoretical operative but as a conceptual transgression.

According to Goldschmidt, micromutation performs the adaptive function within the specious assemblage in a similar manner to that described within *The Origin of Species*. For Darwin, micromutational accumulations were of the gradualist mode and while giving an appearance over time of the transformative ability of the species as a whole, the variation is formally contained within the domain of the species. That is, micromutation is the adaptation *within* the species, *of* the species. For Goldschmidt, however, while subspecies remain as difference *within* a species, "the limit between two species or *rassenkreise* [subspecies] ought to be in the nature of the hiatus, an unbridged cleft" (*MB*, 142). The assertion, thus, of *The Material Basis of Evolution* is that variation occurring *within* a species could not compound into the variation *between* species. It is across the opening, the "unbridged cleft," from which the monster looms forth from Darwin's second "domestication" strategy. Indeed the monster emerges in a manner that resonates with what Braidotti was to describe in philosophy as a process monster and referred to as a "promising monster." "I would like to propose a redefinition, the monster is a process without a stable object. It makes knowledge happen by circulating, sometimes as the irrational non-object."[19] The gradualist appropriation of the monster is destabilized by the dismantlement of the bridge between micro and macromutation, where, according to Goldschmidt:

> (S)ubspecies are actually, therefore, neither incipient species nor models for the origin of species. They are more or less diversified blind alleys within the species. The decisive step in evolution, the first step toward macromutation, the step from one species to another, requires another evolutionary method than that of sheer accumulation of micromutations (*MB*, 183).

Where micromutation is the gradualist slow motion of a species: this "step toward macromutation" is the movement toward a monster. The origin of the incipient species is not shared in the collective of the species but is individuation itself and the body of the *hopeful monster* is demonstrative of new species; where "a monstrosity appearing in a single genetic step might permit the occupation of a new environmental niche and thus produce a new type in one step" (*MB*, 390). The result is that though the *hopeful monster* remains operative within the disciplinary bounds of biology it is no longer located within the bounds of the species.

Following the *modern synthesis* in biology, the term "genotype" refers to an organism's genetic code. The term "phenotype" refers to the bodily expression of that code. The genotype in the discourse of Goldschmidt maintains the metastability that it has within the text of the *modern synthesis*; however, it is also the norm of a monster, where "(t)he genotype is, therefore, the inherited norm of reactivity to the ensemble of conditions which may influence the phenotypic expression" (*MB*, 252). The genotype is to transgress with the monster as the sign of its actualisation. From Darwin to Goldschmidt the genotype as a notion moves from being a code of mutation as a form of specious stability to that of the norm against which to measure the phenotypic expression of the individual organis—monster or otherwise. The genotype within *The Material Basis of Evolution* dictates the normal of the monster: Not in regard to its heredity, as or from its parentage, but in regard to the transgressive norm that it, *itself*, establishes.

THE *HOPEFUL MONSTER* OF ARCHITECTURE

1. In Frascari

The text *Monsters of Architecture* continues to stand adjacent to much of the contemporary discourse concerning the architectural body. The evocation, however, of the concept of monster in Frascari's text is as evasive as the monster itself. In *Monsters of Architecture*, Frascari claims that an anthropomorphic method is inherent to the production of architecture. He posits that this given corporeal presence is initiated at the stage of conception and developed through the practice of architectural drawing. For Frascari, this corporeality must yield an imaginative and "meaningful" subject-image that results from the coalescence of sensation, representation and perception.[20] His account of architecture opens itself to new means of thinking about the interaction of bodies and he assembles a monster as a conceptual persona from which to create and construe the objects of architecture. The monster of Frascari's text is best demonstrated in select small textual morsels (phrases, sentences, paragraphs), such as that relating to the "grotesque body":

> The grotesque body is a body in the act of becoming. It is never finished, never completed; it is continually built, continually created; and it is the principle of others' bodies. The logic of the grotesque image ignores the smooth and impenetrable surface of the neoclassical bodies, and magnifies only excrescences and orifices, which lead into the bodies' depths. The outward and inward details are merged. Moreover, the grotesque body swallows and is swallowed by the world. This takes place in the openings and the boundaries, and the beginning and end are closely linked and interwoven (*MA*, 32).

It is lamentable that the occurrence of the monster, ripe with the potential of what Deleuze calls "decentrings, slips, break ins, secret emissions" and Frascari calls "excrescences and orifices" is not allowed the concrete material and semiotic connections to which it aspires because of the forces to which it is subjected. Though, for Frascari, "the grotesque body swallows and is swallowed by the world," this monster of architecture is substantially underfed (*MA*, 32).

The consuming phenomenology of *Monsters of Architecture* exposes the monster (however fertile in the textual morsel) to the contemporary preoccupation of affirming the health of Vitruvian figurality. It is the transgressive potential of the monster that is in question: The folding of beginning and end into the phenomenological body construct is a loss of transgression in an ambiguous unity of bodily being and subjectivity.[21]

2. Mimesis, Metonym, and Monster

That the monster within *Monsters of Architecture* is not a monster *of* this text is explored by the extraction of the concepts "monster" and "metonym" from the theoretical threat that the text of Frascari represents to the key concepts.

When the genealogy of the Vitruvian man is referenced, it is the consuming and persistent mimetic which is maintained between body and architecture that is fundamental. The analogical force is the line of descent, the genotype, of the Vitruvian man.[22] As the mimetic is reliant upon a teleological construction it is intrinsically anthropocentric. In an architectural visitation of the deliberation between "freeplay and history" described by Derrida, Peter Eisenman asserts that:

> [T]his mimetic condition (that is, architecture as metaphoric representations), despite continuous stylistic changes, remained constant for four hundred years. Then, in the late nineteenth century, with the advent of mass technology and the development of the relativistic human science—biology, sociology, psychology, and anthropology—man could no longer maintain his anthropocentric focus or take for granted his centric position and, correspondingly, the "naturalness" of his social organisations.[23]

The distinction made within *Monsters of Architecture* is a shift from the mimetic understanding of the architecture-body relation to an understanding of architecture as *metonymic* of body. A metonym is the direct replacement of one word or notion with another. It differs from mimesis and its modes (such as metaphor and analogy) in that the mimetic is a direct and contiguous substitution rather than a connection based on similarities or shared traits. When Frascari describes architecture as *metonymic* of body, he is suggesting the indifferentiability of architecture and the body: an embodiment. The potential of the concept to desecrate the Vitruvian genealogy is rich. The removal of the gap or difference that exists between the body and its exterior context represents a deletion of geometric intercession as a possible resolution of the (body/world) dualism. There is a potential for the body to constitute its own norms in relational discourse via the tool of metonym without the intercession of anthropocentric and figural geometries to resolve the dualism.

This would be the case were it not for Frascari's own reconstruction of the metonym directly within the idiom of anthropomorphism. The discursive legacy of Maurice Merleau-Ponty employed toward a profound resolution of Cartesian dualisms within Frascari's text severely curtails the potential of the metonymic operation.[24] The theoretical mode of Frascari's text relies on three criteria that consistently deliver the concepts of metonym and monster back to mimesis and figurality: The *intentionality* of the phenomenological model; the lack of a *real external* to perception; and the deferral to the *preobjective*, confounds the potential of monster and metonym alike.

First, *intentionality* refers to the non-diagrammatic projective or teleological drive. For Merleau-Ponty intentionality refers to "directedness" or "significance": "It is . . . intentional, which means that it does not rest in itself as a thing, but that it is directed and has significance beyond itself" (*PP*, 213). Merleau-Ponty defines intentionality in broader terms as, "the same demand for awareness and the same will to seize the meaning of the world as that meaning comes into being" (*PP*, xxiv). For the monster, the denial of significance "in itself" and the "will to seize meaning" suggest it cannot exist as difference *in itself.*

Second, the absence of a "*real external*" suggests reliance upon internal essence or what Deleuze (and Guattari) refer to as the "subject-thought" with which phenomenology is concerned (*TPC*, 378). Merleau-Ponty argues that *seeing* as a corporal outcome inevitably precedes perception as an interior process.[25] According to him, since immanence always comes from an outside, no rigorous examination of the thing "in itself" will illuminate it; and because immanence is already internal to the thing from the instant of its origin, no quantity of inspection of external conditions (social, cultural, economic, authorial) will bring us any closer to understanding it.

Frascari suggests that the representational space of the architectural image is the space of remote mental inhabitation; it differs from the physical space of the body and can only be conceptualised through the analogical projection of synaesthesia (*MA*, 9–10). This anthropocentric approach attempts to interiorise the visceral and tactile material dimensions of architecture in order to engage the bodily senses as though one were actually physically part of the projected design (what one may have regarded as an exterior). The metonymic relation with the potential to rid the body of the illusions of anthropocentrism and to externalise self is instead engaged internally as a "trope." According to Frascari:

> [T]he role of radical anthropomorphism is to introduce another fertile procedure for the making of architecture. The trope principally used in anthropomorphism is metonymy, a unity of contrasting elements that forms a conventional sequence through which sense is displaced or deferred (*MA*, 7).

Thirdly, to conceive of anthropomorphism devoid of "sense" (that which can be presumed to be consciousness), is not the presentation of an embodiment devoid of the stability and unity that consciousness presupposes (which would consequent a liberation of the concept from anthropomorphic normativity). The "displacement of sense" that Frascari suggests is the *deferral to a pre-objective* primordial relationship we have to our bodies and the world; a coherence that anthropomorphic intentionality presupposes.[26]

For Merleau-Ponty, the self is constructed in terms of the *cogito*.[27] Although Merleau-Ponty's use of *cogito* differs fundamentally from that of Descartes, it remains a "given." Consciousness for Merleau-Ponty is not Cartesian mental matter balanced with, or "contained" in, that which is physical. Rather, Merleau-Ponty proposes human existence in terms of the "lived-body" (*le corps propre*), that is intended to denote a body as it is lived and, unlike a Cartesian material object, the lived-body is a "pre-objective" concrete unity of interdependent physical and psychological characteristics (*PP*, 416). The lived-body is a unity of thought-in-act. It is an organization of powers for interpreting and internalising the world. By this logic, though the *cogito* is translated as "I can" rather than "I think," it remains "I"—an expression of anthropocentric unity. The phenomenological character that Frascari's text assigns the metonym and the monster, removes from that monster its potential to transgress: The monster is contained as external pathology measured against the unitary, stable anthropoid.

CONCLUSION

In the discourse of biological fixism that preceded Darwinism the mutation is considered pathological, but in the discourse of graduated transformism, in

The Origin of the Species, the mutation is sub-pathological (domesticated). It is in the discourse of macromutation in *The Material Basis of Evolution*, that the mutation is potential. It is in this text that the monster is "hopeful." This *hopeful monster* is that which is both its own norm and the enactment of a new normativity. The *hopeful monster* transgresses the "unbridged cleft" that separates not only existent species but also existent species from potential speciation events. This is not merely an affirmation within the discourse of the biological sciences of the normativity of the *hopeful monster*, but significantly a monstering of the given.

In a discourse concerned for the relationship between the body and architecture Frascari suggests that the locality of myth in architecture requires an embodiment as monster:

> The embodiment of myth in architecture serves to reduce . . . the absolutism of reality, creating a breathing space, making a symbolic niche that protects the human animal symbolicus from the fundamental anxiety activated by the relationship between his biological nature and the natural environment (*MA*, 9).

It is as "embodiment" that the monster functions as a relief of the "anxiety" that exists between a human (*animal symbolicus*) as a fixed identity and an exterior. Of concern is that the monster is merely pathology to a given normal man in that the "breathing space" created is not a space of transgression. The monster in the theoretical domestication of phenomenology remains an assertion of the stability of the anthropomorph; swallowed as a test of the health of "given normal men."

The description in the textual morsel, where the "logic of the grotesque *image ignores* the smooth and impenetrable surface of the neoclassical bodies, and magnifies only excrescences and orifices" is a description of a wonderfully monstrous offspring (*MA*, 32; emphasis added). Though the monster may be sincerely domesticated in biological theory, despite Goldschmidt's *Materials for the Study of Evolution*, and may be constructed as an armouring of the body of a normal man in Frascari's *Monsters of Architecture*, it does not discount the assertion that the "image ignores"; and for this we can be hopeful.

NOTES

1. Rosi Braidotti, *Nomadic Subjects: Embodiment and Sexual Difference in Contemporary Feminist Theory* (New York: Columbia University Press, 1994), 100. Hereafter cited as *NSE*.
2. Moira Gatens, *Imaginary Bodies: Ethics, Powers and Corporeality* (London/New York: Routledge, 1996), viii.
3. Gilles Deleuze, *Empiricism and Subjectivity; An Essay on Hume's Theory of Human Nature* (New York: Columbia University Press, 1991), 109.
4. Marco Frascari, *Monsters of Architecture: Anthropomorphism in Architectural Theory* (Lanham, MD: Rowman & Littlefield Publishers, 1991). Hereafter cited as *MA*.

5. Richard Goldschmidt, *The Material Basis of Evolution* (New Haven, CT: Yale University Press, 1940). Hereafter cited as *MB*.

6. Gilles Deleuze, *Negotiations: 1972–1990* (New York: Columbia University Press, 1995), 6. Refer also to David Hume, *A Treatise of Human Nature*, ed. Ernest C. Mossner (London: Penguin, 1985 [1739]), 300.

7. Gilles Deleuze, *Bergsonism*, trans. Hugh Tomlinson and Barbera Habberjam (New York: Zone Books, 1988), 8 in *Le Bergsonisme* (Paris: PUF, 1966).

8. David Hume, *Essays and Treatises on Several Subjects* III, sec. 4, part I, 43–5; sec. 5, 66–7.

9. Gilles Deleuze, *Difference and Repetition* (New York: Columbia University Press, 1994), 222.

10. The first expression of the term "hopeful monster" occurred in Richard Goldschmidt, "Some Aspects of Evolution," *Science* 78 (1933): 539–47.

11. The present chapter differentiates between "heredity" which is a biological line of descent and a "genealogy," which is utilized to describe a discursive connection.

12. Charles Darwin, *On the Origin of Species By Means of Natural Selection, or the Preservation of Favoured Races in the Struggle for Life* (Cambridge, MA: Harvard University Press, 1966), 44. Hereafter cited as *OS*.

13. Gilles Deleuze and Félix Guattari, *A Thousand Plateaus* (Minneapolis: University of-Minnesota Press, 1987), 12. Hereafter cited as *TPC*.

14. Darwin is to concede within the fifth edition of *The Origin of Species* the view that favorable variations in a single individual could not spread through entire populations, in what is a response to a review within the *North British Review* (1867), the author of which was a Scottish engineer, Fleeming Jenkin. As Darwin's concession did not impact upon the argument to be developed by Goldschmidt, so it will not impact upon the present chapter. Reference is made to Stephen Jay Gould, "Fleeming Jenkin Revisited; This Obscure, but Able, Victorian Gentleman Convinced Darwin Himself On an Important Evolutionary Point," *Natural History* 94 (June 1985): 14–19.

15. The compatibility of the notion of "variability" within the discourse of Darwin to the concept of "mutation" (principally as micromutation), within the discourse of mutationism is fostered by the *modern synthesis*.

16. Stephen Jay Gould, "The Hardening of the Modern Synthesis," in *Dimensions of Darwinism*, ed. Marjorie Grene (Cambridge: Cambridge University Press, 1983), 71–93.

17. Peter J. Bowler, *The Eclipse of Darwinism: Anti-Darwinian Evolution Theories in the Decades around 1900* (Baltimore: The John Hopkins University Press, 1983), 178.

18. The enduring nature of the attack of Goldschmidt upon neo-Darwinism is headed by paleo-biologists for whom discontinuity of evolution would deposit significant academic importance upon the fossil record that they preside over. Reference is made to Stephen Jay Gould, "Is a New and General Theory of Evolution Emerging?" *Paleobiology* 6 (1980): 119–30; Stephen Jay Gould, "Return of the Hopeful Monster," *The Panda's Thumb, More Reflections in Natural History* (Hammondsworth: Penguin, 1983), 186–93; Peter Bowler, "The Modern Debates," *Evolution, The History of an Idea* (Berkeley: University of California Press, 1984), 322–6; Michael Ruse, "Is the Theory of Punctuated Equilibria a New Paradigm?" *The Darwinian Paradigm: Essays on its History, Philosophy, and Religious Implications* (London: Routledge, 1993), 118–45.

19. Rosi Braidotti, "Signs of Wonder and Traces of Doubt: On Teratology and Embodied Differences," in *Between Monsters, Goddesses and Cyborgs: Feminist Confrontationswith Science, Medicine and Cyberspace*, eds. Rosi Braidotti and Nina Lykke (London: ZED-Books, 1996), 150.

20. Schawn Jasmann, "Virtual Architecture and the Role of Inscription," in *Hybrid Reality: Art, Technology and the Human Factor*, ed. Hal Thwaites (Montreal: International Society on Virtual Systems and Multimedia, 2003), 422–27.

21. That Frascari predisposes his concept of architectural monster to the further "domestication" of a phenomenological framework is illustrated by an enthusiastic review of the text by Alberto Perez-Gomez; see Alberto Perez-Gomez, "Monsters of Architecture," book review, *Journal of ArchitecturalEducation* 46, no. 1 (September 1992): 60.

22. Robert McAnulty, "Body Troubles," in *Strategies in Architectural Thinking*, eds. John Whiteman et al. (Cambridge, MA: MIT Press, 1992), 180–97; Dalibor Vesely, "The Architectonics of Embodiment," in *Body and Building: Essays on the Changing Relation of Body and Architecture*, eds. George Dodds and Robert Tavernor (Cambridge, MA/London: MIT Press, 2002), 28–43.

23. Peter Eisenman, *Houses of Cards* (New York: Oxford University Press, 1987), 170.

24. Maurice Merleau-Ponty, *Phenomenology of Perception* (New York: Humanities Press,1962). Hereafter cited as *PP*.

25. Maurice Merleau-Ponty, *The Visible and the Invisible* (Evanston: Northwest University Press, 1968), 3–14. Translated by Alphonso Lingis from *Le visible et l'invisible* (Paris: Gallimard, 1964).

26. That the monster of Frascari departs from any other base than that of the anthropomorphic is not given consideration in the text, where it is, further, a regret that the architectural discipline has "forgotten the process of the Vitruvian figurata similitudine" (*MA*, 111).

27. Maurice Merleau-Ponty, *The Primacy of Perception and Other Essays on Phenomenological Psychology, the Philosophy of Art* (Evanston: Northwestern University Press, 1964), 3f.

Chapter Fifteen

Design and New Materialism

Neil Leach

Within contemporary architectural design a significant shift in emphasis can be detected—a move away from an architecture based on purely visual concerns toward an architecture justified by its performance. Structural, constructional, economic, environmental, and other parameters—concerns that were once relegated to a secondary level—have become primary, and are being embraced as positive inputs into the design process from the outset. Architecture, it would seem, is no longer so preoccupied with style and appearance. It is as though a new paradigm has emerged.

This new paradigm can be understood as an attempt to overcome the scenography of postmodernism. It is an attempt to locate architectural discourse within a more objective framework, where efficient use of resources supersedes the aesthetic indulgences of works that came under the broad heading of postmodernism, which might include not only the somewhat conservative movement noted for it decorative use of applied decorative motifs—as postmodernism is understood most commonly within architectural culture—but also more progressive movements such as deconstructivism, all of which privilege appearance over performance.

This is by no means a universal development. Many areas of architectural production remain deeply rooted in postmodern concerns for appearance, and no doubt architectural culture would be poorer if all architects were to subscribe to the same approach. However, it does represent a significant shift in concerns not only in the various "hot-spots" of architectural production—cities such as London, New York, Rotterdam, and Los Angeles—but also in the cities where the designs of various progressive architects from around the world are now being built. The structural logic that informs the "Bird's Nest," "Water Cube," and CCTV Headquarters building in Beijing, no less than the environmental logic and concerns over sustainability that are beginning to inform various developments elsewhere, suggest that this is a global phenomenon.

We might describe this privileging of performance within the design process as an interest in "morphogenesis."[1] Used initially in the realm of biological sciences, the term refers to the logic of form generation and pattern making in an organism through processes of growth and differentiation. More recently it has been appropriated within architectural circles to designate an approach to design that seeks to challenge the hegemony of top-down processes of form making, and replace it with bottom-up logic of form-finding.[2] The emphasis is therefore on material performance over appearance, and on processes over representation.[3]

What we need to recognize, then, is that there might be an apparent formal similarity between the "non-standard" forms of architects such as Frank Gehry and the work of a new generation of architects, such as Atelier Manferdini, Matsys, LAVA, OCEAN, and Material Ecology, but there is an enormous difference in terms of design methodology. Where this new generation exhibits an increasing interest in morphogenetic questions such as performativity and form-finding, Gehry represents a more traditional, postmodern approach toward design, where the architect is perceived as the genius creator who imposes form on the world in a top-down process, and the primary role of the structural engineer is to make possible the fabrication of the designs of the master-architect, as close as possible to his/her initial poetic expression. Meanwhile the more contemporary architects operating within the new morphogenetic paradigm can be seen more as the controllers of processes, who facilitate the emergence of bottom-up form-finding processes that generate structural formations.

The difference then lies in the emphasis on form-finding over form-making, on bottom-up over top-down processes, and on formation rather than form. Indeed the term "form" itself should be relegated to a subsidiary position to the term "formation." Meanwhile "formation" must be recognized as being linked to the terms, "information" and "performance." When architecture is "informed" by performative considerations, it becomes less a consideration of form in and of itself, and more a discourse of material formations. In other words, "form" must be "informed" by considerations of "performative" principles to subscribe to a logic of material "formation."

However, the logic of morphogenesis in architecture is not limited to questions of design methodology. It also extends into an ethical arena. If we can find forms that operate more efficiently from a structural point of view, then we can use fewer materials, equally if we can devise forms that perform more efficiently in terms of energy consumption, we will consume less energy in heating or cooling our buildings. In either case morphogenetic design will help to preserve the world's resources. As such it can be taken not only as a critique of the scenography of postmodernism, but also as an ethical argument about the environment.

MATERIAL COMPUTATION

Biology provides one of the major sources of inspiration for research into morphogenesis in architecture. Nature operates largely through the logic of optimisation, and therefore can offer important lessons for architects. Biomimetics—the study of what we can learn by replicating the mechanisms of nature—has thus emerged as an important field of research. It is not simply that nature can inspire products such as Velcro or recent fabrics used in the manufacture of swimwear that are based on the hydrodynamic properties of shark's skin. Rather nature itself can teach us important lessons about the efficiency of certain structural organizations. Following on from the early experimentation of Antonio Gaudi, Frei Otto has become a champion of observing the behavior of certain structures in nature, and re-applying their principles through analogue modelling. Thus spiders' webs and soap bubbles can provide deep insights into the behavior of form-finding lightweight structures.

These observations come under the heading of material computation. They offer us analogue forms of computation, which—despite the apparent crudeness of the modeling process—are actually highly sophisticated means of understanding structural performance. To describe them as a form of computation is not to undermine the role of digital computation. Rather it is to recognize that computation is everywhere in nature. Computation—a term derived from Latin, "*computare*," to "think together"—refers to any system where individual components are working together. But it is equally important to recognize that digital computation has its limitations. Digital computation necessarily involves the reduction of the world to a limited set of data, which can be simulated digitally, but it can never replicate the complexity of a system such as a soap bubble, whose internal structural computation involves an intricate balance between highly complex surface material organizations and differential atmospheric pressures.

A number of contemporary architects have re-examined the works of Antonio Gaudi and Frei Otto, and found in them sources of inspiration for the new morphogenetic generation of form-finding research, often coupling the lessons of their analogue experimentation with more contemporary digital techniques. Mark Goulthorpe describes his work as a form of "post-Gaudian praxis," while Mark Burry, as architectural consultant for the completion of Gaudi's *Sagrada Familia* church, has been exploring digital techniques for understanding the logic of Gaudi's own highly sophisticated understanding of natural forces. Meanwhile, Lars Spuybroek has performed a number of analogue experimentations inspired by the work of Frei Otto, as a point of departure for some innovative design work that also depends on more recent software developments within the digital realm.[4]

This work points toward a new "performative turn" in architecture, a renewed interest in the principles of structural performance, and in collaborating more empathetically with certain progressive structural engineers. But this concern for performance may extend beyond structural engineering to embrace other constructional discourses, such as environmental, economic, landscaping, or indeed programmatic concerns. In short, what it amounts to is a "folding" of architecture into the other disciplines that define the building industry.[5]

DIGITAL COMPUTATION

Not surprisingly for an age dominated by the computer, this interest in material computation has been matched by an interest in digital computation. Increasingly the performative turn that we have witnessed within architectural design culture is being explored through new digital techniques. These extend from the manipulation and use of form-generating programs from L-Systems to cellular automata, genetic algorithms and multi-agent systems that have been used by progressive designers to breed a new generation of forms, to the use of the computer to understand, test out and evaluate already designed structures.

The seemingly paradoxical use of the immaterial domain of the computer to understand the material properties of architecture has spawned a new term in architecture, "digital tectonics." In other words the old opposition between the highly material world of the tectonic, and the immaterial world of the digital has broken down. What we have instead is a new tectonics of the digital or "digital tectonics."[6]

A certain genealogy can be detected in the use of the computer in architecture. What distinguishes this new digital paradigm from early uses of the computer in the architectural arena, is that it reinterprets the computer not simply as a sophisticated drafting tool—an extension, in other words, of the possibilities of the previous paradigm of ink on tracing paper logic—but also as a device that might become part of the design process itself. With this we see a development in the very nature of the architect from the demiurgic "form-giver," who in Alberti's terms, "imagined in the mind, and realised though construction," to the architect as the controller of generative processes, where the final appearance is a product not of the architect's imagination alone, but of the generative capacities of computer programs. It is not that the architect here is any less imaginative. Rather the architectural imagination has been displaced into a different arena—into the imaginative use of various processes.[7]

But even within the logic of digital tectonics there is a certain genealogy of development. Computational methodology had first been used as a means of testing and thereby verifying and supporting the initial designs of the architect. The objective here was simply to use the computer to make the designs of the architect realisable. The only significant contribution to the design process occurred when findings of this process influenced the original design and forced minor amendments to that design. Examples here would include the use of software to test out the acoustic performance of the Greater London Authority building by Foster and Associates.[8]

Also, the computer could make occasionally a more precise, structural definition of a loosely formulated architectural concept. Examples here would include the use of algorithms by Chris Williams to define the form of the glass canopy to the British Library, and "dynamic relaxation technique" to define the precise vectorial layout of the mullion system.[9]

A second generation of computational methodology, however, can be detected in the work of Kristina Shea, whose eifForm program serves to generate structural forms following a stochastic, non-monotonic method using a process of structural shape annealing.[10] The "designer" merely establishes certain defining coordinates and then unleashes the program, which eventually "crystallises" and resolves itself into a certain configuration. Each configuration is a structural form, which will support itself against gravity and other prescribed loadings, and yet each configuration thrown up by the program is different. Such is the logic of a bottom-up, stochastic method.

It is programs such as this that reveal the true potential of the digital realm in influencing the process of design itself, by opening up fields of possibilities. The computer, then, emerges not only as a prosthetic device that extends the range of the architectural imagination, but also—much like a calculator—as a tool of optimization that offers a more rigorous means of searching out possible options other than what could be described as the pseudo-computational logic often dominating contemporary practice.

DELEUZE AND NEW SCIENTIFIC THINKING

A similar shift can be detected within architectural theory. If during the 1980s and 1990s architectural theory was dominated by an interest in literary theory and continental philosophy—from the structuralist logic that informed the early postmodernist quest for semiological concerns in architectural writers such as Charles Jencks and Robert Venturi to the poststructuralist enquiries into meaning in the work of Jacques Derrida, that informed the work of architects such as Peter Eisenman and Bernard Tschumi—the first decade of the twenty-first century has been characterized by a waning of interest in this branch of theory.

This is not to endorse the position of architectural theorist, Michael Speaks, who claims that we have witnessed the "death of theory."[11] For such a theory, it could be argued, is merely an anti-theory theory in that there is surely no position that stands outside theory. Any form of practice must be informed by a theoretical impulse, even if it is a positivistic one that purportedly disdains theory. Rather, I would claim, what we are witnessing is the ascendancy of a new branch of theory, one that engages with science, technology, and material behavior.

Much of this new theoretical work finds its grounding in the thinking of Gilles Deleuze. For if there is one continental philosopher of the twentieth century who has survived the shifting sands of intellectual fashion, where the spotlight has moved on from linguistic concerns toward a more material understanding of the world, it is Deleuze, who has become the philosopher of choice within certain progressive architectural circles, where the concept of the diagram holds a dominant position, and where questions of material performance have become paramount.

Deleuze makes few explicit references to architecture in his writings, but in *A Thousand Plateaus*—which he co-wrote with Félix Guattari—there is a very precise formulation offered about two alternative sensibilities toward architectural design (*TPC*). It is as though the whole history of architecture can be divided into two contrasting yet reciprocally related outlooks. One would be a broadly aesthetic outlook that tends to impose form on building materials according to some preordained "template." (Here one immediately thinks of the role of proportions and other systems of visual ordering.) The other would be a broadly structural outlook that tends to allow forms to "emerge" according to certain programmatic requirements.

Deleuze and Guattari describe the first sensibility as the "Romanesque." The term seems somewhat restrictive, in that the principle covers a range of stylistic approaches, which broadly come under the umbrella of the Classical. This would include not only the Classical as such—the Roman and Greek styles which mutated through the Romanesque, into the Renaissance, Mannerism, Baroque, and Neo-Classical—but also any outlook, which focuses on appearance rather than performance.

The second could be broadly defined as the Gothic, which is configured not as a style as it was in the nineteenth century, but as a method. It is a way of designing that privileges process over appearance. Form "emerges" with time, much as the Gothic vault evolved over the centuries, becoming ever more refined in its structural efficiency, until it reached such intricacies as fan vaulting. Within this outlook architecture becomes the result of competing forces, a programmatic architecture that registers the impulses of human habitation, and adapts to those impulses. Deleuze and Guattari analyze the distinction between the Gothic spirit and the Romanesque as a "qualitative" distinction, between a static and dynamic model of understanding architecture.[12]

Rather than describing these two different outlooks in terms of style, Deleuze and Guattari refer to them in terms of different "sciences." One is a science of intensive thinking that perceives the world in terms of forces, flows, and process.[13] The other is a science of extensive thinking that seeks to understand the world in terms of laws, fixity, and representation. In other words, the one is a smooth science, and the other striated. Deleuze and Guattari also describe this opposition as being that between a nomad, war-machine science and a royal, state science. The latter is a science of fixed rules and given forms, a hierarchical system imposed from above.[14] By contrast, the nomad war-machine science is a bottom-up model that responds in each individual instance to the particularities of the moment.[15] It is this Gothic spirit that is seemingly celebrated by certain contemporary architects working under the aegis of Deleuze's thinking in this "performative turn" within architectural culture. Out of Deleuze's thinking a new performative theory of architecture has emerged.

NEW MATERIALISM

I will call this new theory, "New Materialism," a term coined by Manuel DeLanda, a self-styled "street philosopher" who has developed a certain reputation for his interpretation of the work of Deleuze, and who has had a major impact on architectural thinking through various teaching positions he has held in architectural schools in East Coast America. DeLanda uses this term to define a new theoretical paradigm, which operates as a retrospective manifesto for a movement whose genealogy stretches back to the work of biologist D'Arcy Wentworth Thompson, philosopher Henri Bergson, and beyond, but also incorporates much recent scientific thinking that has emerged from centres of interdisciplinary scientific research, such as the MIT Media Lab and the Santa Fe Institute. New Materialism has yet to be defined in concrete terms even as a philosophical concept. Indeed if we are to look for a definition of the term, the best we could do is to see it articulated indirectly through DeLanda's own writings, such as *A Thousand Years of Nonlinear History* in which he recasts the whole history of urban growth within a framework of material processes, and Intensive Science and Virtual Philosophy, which examines the role of scientific theory in Deleuze's writing.[16]

The key behind New Materialism is to recognize that the emphasis today should not be on symbols but on material expressions. It is as though the postmodern "linguistic turn" that constructed an elaborate intellectual artifice out of the quest for meaning, so that all cultural artefacts could be "read" in semantic terms, could be faulted precisely because of its urge to understand the world in terms of meaning. The quest for meaning itself begets meaning. As this linguistic turn began to unravel—as poststructuralist theorists questioned the flexibility of structuralist univalent formulations of the relationship between signified and signifier—it slowly became clear that even this quest for meaning was a flawed intellectual enterprise. Instead of unfolding the logic of communicative discourse, poststructuralism painted itself into the intellectual corner of challenging the very epistemological basis of how we might ever know the "other." It is as though the ultimate contribution of poststructuralism was to reveal the necessity of problematization itself. Yet this led to another impasse. If we are always trapped by the double bind of never knowing the other, the best we can hope for is not to extract ourselves from this problem, but rather to recognize the problem in the first place. Once we are aware of a problem it becomes a different kind of problem—not one by which we are trapped, but one with which we can begin to deal. And yet poststructuralism itself can offer us no instructions as to how to deal with the actual problem.

Jean-François Lyotard provides something of an intellectual rationale for this shift away from a quest for meaning through his concept of the "differend," which exposes the limit case of meaning. With reference to those events that defy any meaningful explanation, such as the Holocaust, Lyotard refers to the concept in terms of the "incommensurate" or that which cannot be expressed in language. As Lyotard writes, "The differend is the unstable state and instant of language wherein something which must be able to be put into phrases cannot yet be."[17] Yet what New Materialism challenges is not the failure or limits of meaning, but the very urge to seek meaning, as though each action or item is to be justified in terms of its relation to meaning.

New Materialism can be compared and contrasted with the old Historical Materialism of Karl Marx. Famously, Marx had turned Hegelian dialectics "on its head," and—against Hegel's idealistic theory of the dialectic—had stressed the primacy of the material world. Equally there are echoes of Marx's famous dictum from his Theses on Feuerbach, "The philosophers have only interpreted the world, in various ways; the point is to change it," in relation to DeLanda's critique of postmodern hermeneutics. There are echoes too of Marx's basic premise that what we see on the surface of cultural phenomena is the product of deeper underlying forces. But New Materialism extends the range of Historical Materialism. For Marx the only form of economic production considered was labor, whereas for New Materialism any cultural expression—social, economic, or political—can be understood in terms of the forces that produce it.

NEW MATERIALISM IN ARCHITECTURE

DeLanda has also written a series of articles drawing upon Deleuze's notion of the "Gothic" spirit, and exploring its relevance for thinking in terms of material behavior.[18] Most recently he has published a series of articles on New Materialism in Domus, looking at biomimetics, intelligent materials and other contemporary material concerns. However, although his work spans the two diverse—and yet paradoxically related—areas of urban formation and materiality, DeLanda has seldom addressed architectural design itself. Nonetheless, a clear sense of the relevance of DeLanda's writings for architecture can be gleaned in particular from his seminal book on urban growth, *A Thousand Years of Non-Linear History*.

Here DeLanda stresses the internal morphogenetic forces that produce society within the respective domains of economics, biology, and linguistics. For DeLanda we need to revise our understanding of history to see it as the interface between the self-organizing potential of matter and energy set against the whim and will of human history. For DeLanda cities emerge as a form of "exoskeleton" to human operations, which—like the internal "endoskeleton" of the human skeleton—was a mineralization of deposits that served to constrain the movement of not only human flesh, but also goods. These cities arise "from the flow of matter-energy" through society. DeLanda then goes on to explore how crucial moments in history have led to "phase transitions" prompted by crucial social developments, often prompted by new technological revolutions, such as the development of agriculture or the Industrial Revolution. These intensify the flow of energy, although the actual forms cities take is dependent on human decision-making, through markets and bureaucracies composed of "self-organized meshworks of diverse elements, versus hierarchies of uniform elements" (*TYN*, 32).

This thinking is echoed in the other theorists, such as Michael Weinstock who likewise describes the city in material terms: "City forms are material constructs that are composed of a spatial array of dwellings, a pattern of streets and public spaces together with differentiated buildings of varying sizes associated with the regulation of energy and material flow, and the extension of the metabolic network across the surrounding territory."[19] These operations become even more complex at the level of the metropolis: "Metropolitan or "mother" cities developed increasingly complex information systems that in turn enabled the further development of systematic transformations of materials for the construction of artifacts, buildings, and cities. Information flowed back to the colonies, accelerating their local expansion and increasing their complexity in turn. . . . The emergence and subsequent evolutionary development of information systems and the systems of cities were strongly coupled, each acting as positive feedback on the expansion and growth in complexity of the other" (*AE*, 216–17).

In urban terms, DeLanda has delineated a new understanding of material growth—one that focuses less on the precise design of physical artefacts—the paraphernalia of the city itself, its parks, buildings, and all that generally comes under the definition of "Urban Design"—and more on the underlying forces that influence its growth—the more abstract economic forces and policy making processes that come under the definition of "Urban Planning."

In architectural terms, then, the message is that we should be concerned less and less with symbolic content—what a building might "mean"—and more and more with performance and material behaviors. Just as, in DeLanda's terms, we need to understand our cities in terms of the economic, social, and political forces that generate them, so too we need to understand architectural design in terms of material processes.[20]

Within this new configuration the economist, the scientist and the engineer are among the reassessed heroes of our intellectual horizon, and figures such as Cecil Balmond, Hanif Kara, and Mutsuro Sasaki have become the new "material philosophers"—to use another term adopted by DeLanda—of New Materialism. But it is not just materialist philosophies that have seized the imagination of architectural theorists. So too, scientific thinking itself has begun to find its place in the architectural curriculum, from the early observations of D'Arcy Thompson on growth and form to more recent theories, such as "emergence," popularized by Steven Johnson, and Stephen Wolfram's discourse of "A New Kind of Science," both of which deal with complexity emerging from a simple set of initial rules.[21] This can be read as a highly positive development within architectural circles in that the domains of science and technology, for so long neglected at the expense of history and theory and treated as largely positivistic domains, have now been re-appropriated and recognized as offering a highly relevant and rich domain of intellectual enquiry.

If we add to these the developing interest in computational methodology—the possibility of scripting, parametric modelling, and performance-based generative techniques such as multi-agent systems or genetic algorithms, but equally too new digital fabrication processes, such as CNC milling, laser cutting, 3D printing—we can begin to define a broad shift that has already appeared in certain progressive architectural circles, and that is beginning to spread into mainstream architectural culture.

In short, whether we are to look at the techniques of production itself, and the increasing importance of the digital realm in both fabrication and design, or at the new sensibilities that are informing the politics of production, we can detect a new approach to architectural design. What we are witnessing, I would argue, is not only a new generation of architectural designs, but also a new theoretical paradigm.

Welcome to the architecture of New Materialism.

NOTES

1. Morphogenesis is derived from the Greek terms, "morphe" (shape/form) and "genesis" (creation).

2. On this see *EMD*, *TTM*.

3. As Achim Menges comments: "Architecture as a material practice is mainly based on design approaches that are characterised by a hierarchical relationship that prioritises the generation of form over its subsequent materialization. Equipped with representational tools intended for explicit, scalar geometric descriptions, the architect creates a scheme through a range of design criteria that leave the inherent morphological and material capacities of the employed material systems largely unconsidered. Ways of materialization, production, and construction are strategized and devised as top-down engineered, material solutions only after defining the shape of the building and the location of tectonic elements. . . . An alternative morphological approach to architectural design entails unfolding morphological complexity and performative capacity from material constituents without differentiating between formation and materialization processes." Achim Menges, "Polymorphism" (*TTM*. 79).

4. See Mark Burry, "Virtually Gaudi"; Mark Goulthorpe, "Gaudi's Hanging Presence"; Lars Spuybroek, "Softoffice" in *Digital Tectonics*, eds. Neil Leach, David Turnbull, and Chris Williams (London: Wiley, 2004).

5. As Alejandro Zaera-Polo and Farshid Moussavi comment, their interest is to recognize the other disciplines in the building industry not simply as offering a service that should be treated as an afterthought in the design process, but rather an important range of design considerations that should be embraced and incorporated into the early stages of the design process itself. Farshid Moussavi and Alejandro Zaera-Polo (Foreign Office Architects), "Rollercoaster Construction" in *Designing for a Digital World*, ed. Neil Leach (London: Wiley, 2002), 80–87.

6. On this see "Introduction" in "Design by Algorithm" in *Digital Tectonics*, eds. Neil Leach et al. (London: Wiley, 2004), 4–12.

7. On this see Manuel DeLanda, "Deleuze and the Use of the Genetic Algorithm in Architecture" in *Designing for a Digital World*, ed. Neil Leach (London: Wiley, 2003), 117–20.

8. On this see Michael Weinstock and Nikolaos Stathopoulos, "Advanced Simulation in Design" in *TTM*, 56.

9. On this see Chris Williams, "Design by Algorithm" in *Digital Tectonics*, 78–85.

10. Annealing refers to the method of heating and cooling metals. The eifForm program simulates this process, so that the eventual form "crystallises." The process is stochastic because it contains a random element to the search process, which is controlled to allow for exploration of concepts that are initially worse than the current design. It is therefore also non-monotonic, in that it is constantly under revision, often negating previous developments. For a discussion of the eifForm program see Kristina Shea, "Creating Synthesis Partners" in *Contemporary Techniques in Architecture*, *AD* 72 (2002): 42–5.

11. Michael Speaks, "No Hope, No Fear" in *ARQ* 6, no. 3 (2002): 209–12.

12. "Gothic architecture is indeed inseparable from a will to build churches longer and taller than the Romanesque churches. Ever further, ever higher. . . . But this difference is not simply quantitative; it marks a qualitative change: the static relation, form-matter, tends to fade into the background in favor of a dynamic relation, material-forces. It is the cutting of stone that turns it into material capable of holding and coordinating forces of thrust, and of constructing ever higher and longer vaults. The vault is no longer a form but the line of continuous variation of the stones. It is as if Gothic conquered a smooth space, while Romanesque remained partially within a striated space (in which the vault depends on the juxtaposition of parallel pillars) (*TPC*, 364).

13. "One does not represent, one engenders and traverses. This science is characterized less by the absence of equations than by the very different role they play: instead of being good forms absolutely that organize matter, they are 'generated' as 'forces of thrust' (*poussées*) by the material, in a qualitative calculus of the optimum" (*TPC*, 364).

14. Royal, or State, science only tolerates and appropriates stone cutting by means of *templates* (the opposite of squaring), under conditions that restore the primacy of the fixed model of form, mathematical figures, and measurement (*TPC*, 365).

15. A further way to distinguish these two models of operation is the distinction Deleuze and Guattari make between "minor" and "major" sciences: "the tendency of the broken line to become a curve, a whole operative geometry of the trait and movement, as pragmatic science of placings-in-variation that operates in a different manner than the royal or major science of Euclid's invariants and travels a long history of suspicion and even repression" (*TPC*, 109).

16. Manuel DeLanda, *A Thousand Years of Nonlinear History* (New York: Zone Books/ Swerve Editions, 1997); Manuel DeLanda, *Intensive Science and Virtual Philosophy* (London/ New York: Continuum, 2002). Hereafter cited as *TYN*.

17. Jean-François Lyotard, *The Differend: Phrases in Dispute* (Minnesota: University of Minnesota Press, 1988), 13.

18. See, for example, DeLanda "Material Complexity" in *Digital Tectonics*, eds. Neil Leach, David Turnbull, and Chris Williams (London: Wiley, 2004).

19. Michael Weinstock, *The Architecture of Emergence: The Evolution of Form in Nature and Civilisation* (London: Wiley, 2010), 202. Hereafter cited as *AE*.

20. If, for example, we were to look for an illustration of this new approach in terms of design processes, we might look to the example of stones on a riverbed in some mountain valley. It is not as though the stones collected there were arranged by God—as if s/he had spent an afternoon gardening there and had arranged the stones in a certain way—but by the forces of nature itself. The position of each stone is defined by its shape, weight, and the forces that washed it there after the melting snows create a torrent of water that swept down the mountain.

21. D'Arcy Wentworth Thompson, *On Growth and Form* (New York: Dover Publications, 1992); Steven Johnson, *Emergence: The Connected Lives of Ants, Brains, Cities and Software* (London: Penguin, 2001); Stephen Wolfram, *A New Kind of Science* (London: Wolfram Media, 2002). On Emergence, see also Eric Bonabeau, Marco Dorigo, and Guy Theraulaz, *Swarm Intelligence: From Natural to Artificial Systems* (New York/Oxford: Oxford University Press, 1999); John Holland, *Emergence: From Chaos to Order* (Oxford: Oxford University Press, 1999).

Bibliography

Aarseth, Espen J. *Cybertext: Perspectives on Ergodic Literature*. Baltimore: Johns Hopkins University Press, 1997.

Agamben, Giorgio. *The Coming Community*. Translated by Michael Hardt. Minneapolis/ London: University of Minnesota Press, 1993.

———. "Beyond Human Rights." In Eduardo Cadava, Aaron Levy, and Giorgio Agamben (eds.), *Cities without Citizens*, 3–19. Philadelphia with the Rosenbach Museum and Library: Slought Books, Theory Series no. 1, 2003.

Agrest, Diana I. "The Misfortunes of Theory." In *Architecture from Without: Theoretical Framings for a Critical Practice*. Cambridge, MA: MIT Press, 1991.

Allen, Lindy. "Photographer of Brilliance." In Bruce Rigsby and Nicolas Peterson (eds.), *Donald Thomson: The Man and the Scholar*, 45–62. Canberra: The Academy of the Social Sciences in Australia, 2005.

———. "Tons and Tons of Valuable Material." In Nicolas Peterson, Lindy Allen, Louise Hamby, and Museum Victoria (eds.), *The Makers and Making of Indigenous Australian Museum Collections*, 387–419. Melbourne: Melbourne University Press, 2008.

Archer, Bruce. "A View of the Nature of Design Research." In Robin Jacques and James Powell (eds.), *Design:Science:Method*, proceedings of the *1980 Design Research Society Conference*, 30–47. Guildford: Westbury House, 1981

Arendt, Hannah. *The Human Condition*. Chicago: University of Chicago Press, 1958

Arendt, Hannah, ed. *Illuminations*. Translated by Harry Zohn. London: Fontana, 1973.

Assmann, Aleida. "On the (In)compatibility of Guilt and Suffering in German Memory." *German Life and Letters* 59, no. 2 (April 2006): 187–200.

Augé, Marc. *Non-places: Introduction to an Anthropology of Supermodernity*. London: Verso, 1995.

Baer, Ulrich. "To Give Memory a Place: Holocaust Photography and the Landscape Tradition." *Representations* 69, Special Issue: Grounds for Remembering (Winter 2000): 38–62.

Bal, Mieke. *Narratology: Introduction to the Theory of Narrative*. Toronto: University of Toronto Press, 1997.

Barthes, Roland. "Introduction to the Structural Analysis of Narratives." In *Image-Music- Text.* Translated by Stephen Heath. Glasgow: Fontana, 1977, 79–124.

———. *Camera Lucida: Reflections on Photography*. Translated by Richard Howard. New York: Hill and Wang, 1981; London: Vintage, 2000.

———. *The Responsibility of Forms: Critical Essays in Music, Art and Representation*. Translated by Richard Howard. Berkeley: University of California Press, 1991.

———. *Elements of Semiology*. Translated by Annette Lavers and Colin Smith. New York: Hill and Wang, 1995.

Baudrillard, Jean. *For a Critique of the Political Economy of the Sign*. Translated by Charles Levin. St. Louis: Telos Press, 1981.

———. *Simulations*. New York: Semiotext(e), 1983.

———. *Fatal Strategies*. Edited by Jim Fleming. Translated by Philip Beitchman and W. G. J. Niesluchowski. New York: Semiotext(e), 1990.

———. *Simulacra and Simulation*. Michigan: Michigan University Press, 1994.

———. *The Consumer Society: Myths and Structures*. London: Sage, 1998.

———. *Symbolic Exchange and Death*. Translated by I. H. Grant. London: Sage, 1983.

———. *Screened Out*. New York: Verso, 2002.

Beck, Ulrich, Anthony Giddens, and Lash Scott. *Reflexive Modernization: Politics, Tradition and Aesthetics in the Modern Social World*. Cambridge: Polity Press, 1994.

Beilby, Raymond, and Cecil Hadgraft. *Ada Cambridge, Tasma and Rosa Praed*. Melbourne: Oxford University Press, 1979.

Benjamin, Andrew, ed. *The Lyotard Reader*. Oxford: Blackwell, 1989.

Benjamin, Andrew. *Art, Mimesis and the Avant-Garde: Aspects of a Philosophy of Difference*. London: Routledge, 1991.

Benjamin, Walter. "On Language as Such and on the Language of Man." In Peter Demetz (ed.), *Reflections: Essays, Aphorisms, Autobiographical Writings*. New York: Schocken Books, 1986.

Benyus, Janine M. *Biomimicry: Innovation Inspired by Nature*. New York: Harper Collins, 1997.

Blanchot, Maurice. *The Unavowable Community*. Translated by Pierre Joris. Barrytown: Station Hill Press, 1988.

———. *The Step Not Beyond*. Translated by Lycette Nelson. Albany: SUNY Press, 1992.

———. *The Infinite Conversation*. Translated by Susan Hanson. Minneapolis/London: University of Minnesota Press, 1993.

———. *The Space of Literature*. Translated by Ann Smock. Lincoln/London: University of Nebraska Press, 1993.

———. *The Writing of the Disaster*. Translated by Ann Smock. Lincoln: University of Nebraska Press, 1995/1980.

———. *The Station Hill Blanchot Reader, Fiction & Literary Essays*. Edited by George Quasha. Translated by Lydia Davis, Paul Auster and Robert Lamberton. Barrytown: Station Hill, 1999.

———. *Instant of My Death*. Translated by Elizabeth Rottenberg. Stanford: Stanford University Press, 2000.

Bolter, Jay, and Richard Grusin. *Understanding New Media*. Cambridge, MA: MIT Press, 2000.

Bonabeau, Eric, Marco Dorigo, and Guy Theraulaz. *Swarm Intelligence: From Natural to Artificial Systems*. New York/Oxford: Oxford University Press, 1999.

Bortoft, Henri. *The Wholeness of Nature: Goethe's Way of Science*. New York: Floris Books, 1996.

Bourdieu, Pierre. *Outline of a Theory of Practice*. Translated by Richard Nice. Cambridge: Cambridge University Press, 1977.

———. *The Logic of Practice*. Translated by Richard Nice. Cambridge: Polity Press, 1990.

Bourriaud, Nicolas. *Postproduction: Culture as Screenplay: How Art Reprograms the World*. Translated by Jeanine Herman. New York: Lukas & Sternberg, 2002.

Braidotti, Rosi. *Nomadic Subjects: Embodiment and Sexual Difference in Contemporary Feminist Theory*. New York: Columbia University Press, 1994.

Briggs, Asa. *Victorian Cities*. Harmondsworth, Middlesex, England: Penguin, 1985.

Briggs, Helen."Machines To Match Man by 2029." BBC Science Reporter, BBC News. Last modified February 16, 2008. http://news.bbc.co.uk/2/hi/americas/7248875.stm.

Bruns, Alex. *Blogs, Wikipedia, Second Life and Beyond: From Production to Produsage*. New York: Peter Lang, 2008.

Burchell, Graham, Colin Gordon, and Peter Miller, eds. *The Foucault Effect: Studies in Governmentality*. Chicago: University of Chicago Press, 1991.

Calvino, Italo. *Invisible Cities*. London: Picador, 1979.

Cambridge, Ada. *A Woman's Friendship*. Edited by Elizabeth Morrison. Kensington: New South Wales University Press, 1988.

———. *Thirty Years in Australia*. Ithaca: Cornell University Library, 2009.

Canguilhem, Georges. *The Vital Rationalist*. Edited by Franscois Delaporte. New York: Zone Books, 2000.

Caputo, John, and Mark Yount. *Foucault and the Critique of Institutions*. Pennsylvania: Pennsylvania State University Press, 1993.

Carey, John. *Eyewitness to History*. New York: Avon Books, 1987.

Carpo, Mario. "Tempest in a Teapot." *Log* 6 (2005): 99–106.

Cartwright, William. "Art and Cartographic Communication." In W. Cartwright, G. Gartner, and A. Lehn (eds.), *Art and Cartography*, 19. Berlin Heidelberg: Springer-Verlag, 2009.

Casey, Edward. S. *The Fate of Place: A Philosophical History*. Berkeley: University of California Press, 1997.

———. "Body, Self, and Landscape: A Geo-Philosophical Inquiry into the Place-World." In Paul C. Adams, Steven D. Hoelscher, and Karen E. Till (eds.), *Textures of Place: Exploring Humanist Geography*, 403–25. Minneapolis: University of Minnesota Press, 2001.

Castner, Henry W. "Might There Be a Suzuki Method in Cartographic Education?" *Cartographica* 18, no. 1 (1981): 59–67.

Certeau, Michel de. *The Practice of Everyday Life*. Berkeley: University of California Press, 1984.

Charvet, P. E., trans. *Charles Baudelaire: Selected Writings on Art and Literature*. New York: Viking, 1972, 395–422, in "The Painter of Modern Life. Charles Baudelaire (1863)," Modernism och postmodernism, last modified December 22, 2010, accessed November 15, 2009, http://www.idehist.uu.se/distans/ilmh/pm/baudelaire-painter.htm.

Chase, Athol. "Thomson Time." *Aboriginal History* 3 (1979): 109–10.

Chen, Qiulin. "Salvaged from Ruins." In the proceedings of *The 6th Asia Pacific Triennial of Contemporary Art*, 75–6. Brisbane: Queensland Art Gallery, 2009.

Colomina, Beatriz. *Privacy and Publicity: Modern Architecture as Mass Media*. Cambridge, MA: MIT Press, 1994.

Crary, Jonathan, and Sanford Kwinter, eds. *Zone 6: Incorporations*. New York: Zone Books, 1992.

Cross, Nigel. *Designerly Ways of Knowing*. London: Springer Verlag, 2006.

Darwin, Charles. *On the Origin of Species By Means of Natural Selection, or the Preservation of Favoured Races in the Struggle for Life*. Cambridge, MA: Harvard University Press, 1966.

Davis, Therese. "Remembering Our Ancestors: Cross-Cultural Collaborations and the Mediation of Aboriginal Culture and History in *Ten Canoes*." *Studies in Australasian Cinema* 1 (2007): 5–14.

Davison, Graeme. *The Rise and Fall of Marvellous Melbourne*. Melbourne: Melbourne University Press, 1981.

Deger, Jennifer. *Shimmering Screens: Making Media in an Aboriginal Community*. Minneapolis: University of Minnesota Press, 2006.

DeLanda, Manuel. *A Thousand Years of Nonlinear History*. New York: Zone Books/Swerve Editions, 1997.

Deleuze, Gilles. *Nietzsche and Philosophy*. Translated by Hugh Tomlinson. New York: Columbia University Press, 1983.

———. *Spinoza: Practical Philosophy*. Translated by Robert Hurley. San Francisco: City Lights Books, 1988.

———. *Empiricism and Subjectivity: An Essay on Hume's Theory of Human Nature*. New York: Columbia University Press, 1991.

———. *Difference and Repetition*. Translated by Paul Patton. New York: Columbia University Press, 1994.

———. "What Children Say." In *Essays Critical and Clinical*. Translated by Daniel W. Smith and Michael A. Greco. London: Verso, 1998.

Deleuze, Gilles, and Félix Guattari. *Anti-Oedipus: Capitalism and Schizophrenia*. Translated by Robert Hurley, Mark Seem, and Helen Lane. New York: Viking, 1977; Minneapolis: University of Minnesota Press, 1983.

———. *A Thousand Plateaus: Capitalism and Schizophrenia*. Translated by Brian Massumi. Minneapolis: University of Minnesota Press, 1987; London: Athlone, 1988.

———. *What Is Philosophy?* Translated by Graham Burchell and Hugh Tomlinson. London: Verso, 1994.

———. *Intensive Science and Virtual Philosophy*. London/ New York: Continuum, 2002.

Demetz, Peter, ed. "The Destructive Character." In *Reflections: Essays, Aphorisms, Autobiographical Writings*, 301–3. New York: Schocken Books, 1986.

Derrida, Jacques. *Writing and Difference*. Translated by Alan Bass. Chicago: University of Chicago Press, 1978.

———. "Structure, Sign, and Play in the Discourse of the Human Sciences." In *Writing and Difference*. Translated by Alan Bass. Chicago: University of Chicago Press, 1978.

———. "Living On/Borderlines." In *Deconstruction and Criticism*, 75–176. New York: Seabury Press, 1979.

———. *Positions*. Translated by Alan Bass. Chicago: University of Chicago Press, 1981.

———. *Dissemination*. Translated by Barbara Johnson. Chicago: University of Chicago Press, 1981.

———. "Différance." In *Margins of Philosophy*. Translated by Alan Bass, 1–27. Sussex: The Harvester Press, 1982.

———. *Margins of Philosophy*. Translated by Alan Bass. Sussex: The Harvester Press, 1982.

———. *The Truth in Painting*. Translated by Geoffrey Bennington and Ian McLeod. Chicago: University of Chicago Press, 1987.

———. *Acts of Literature*. Edited by Derek Attridge. Translated by Christine Roulston, 181–220. London/New York: Routledge, 1992.

———. *Aporias*. Translated by Thomas Dutoit. Stanford: Stanford University Press, 1993.

———. *Points . . . Interviews, 1974-1994*. Edited by Elisabeth Weber. Translated by Peggy Kamuf. Stanford: Stanford University Press, 1995.

———. *On The Name*. Edited by Thomas Dutoit. Translated by David Wood, John P. Leavy, and Ian McLeod. Stanford: Stanford University Press, 1995.

———. *Of Grammatology*. Translated by Gayatri Chakravorty Spivak. Baltimore/London: The John Hopkins University Press, 1998.

Dodds, Chris. "Avatars and the Invisible Omniscience: The Panoptical Model within Virtual Worlds." Master's dissertation, RMIT University, School of Creative Media, 2007. http://www.iconinc.com.au//christo/c.dodds_exegesis.pdf.

Dovey, Kim. *Fluid City: Transforming Melbourne's Urban Waterfront*. Oxfordshire: Routledge; Sydney: University of NSW Press, 2005.

Eco, Umberto. *The Role of the Reader: Explorations in the Semiotics of Texts*. Bloomington: Indiana University Press, 1979.

———. *The Open Work*. Translated by Anna Cancogni. Cambridge, MA: Harvard University Press, 1989.

Edquist, Harriet. *Pioneers of Modernism: The Arts and Crafts Movement in Australia*. Melbourne: Miegunyah Press, 2008.

———. "'Precise and Imprecise Geographies' in Christina Stead's Seven Poor Men of Sydney." *The Cartographic Journal* 54, no. 4 (November 1, 2009): 343.

Eisenman, Peter. *Houses of Cards*. New York: Oxford University Press, 1987.

———. "Text as Zero Or: The Destruction of Narrative." In *Re-working Eisenman*, 40–3. London: Academy Editions; Berlin: Ernst & Sohn, 1993.

———. *Written into the Void: Selected Writing 1990-2004*. New Haven, CT: Yale University Press, 2007.

Eleftheriades, Michael. "Architecture or Cinema?: Digital 3D Design and the World of Multimedia." In François Penz and Maureen Thomas (eds.), *Cinema and Architecture: Méliès, Mallet-Stevens, Multimedia*, 138–43. London: British Film Institute, 1997.

Everett, Anna, and John Caldwell, eds. *New Media, Theories and Practices of Digitextuality*. New York: Routledge, 2003.

Featherstone, Mike. *Undoing Culture: Globalization, Postmodernism and Identity*. London: Sage, 1995.

Florida, Richard. *Cities and the Creative Class*. London/New York: Routledge, 2005.

Foray, Dominique. *The Economics of Knowledge*. Cambridge, MA/London: MIT Press, 2004.

Forty, Adrian. *Words and Buildings: A Vocabulary of Modern Architecture*. New York: Thames & Hudson, 2000.

Foster, Hal. "The Archive without Museums." *October* 77 (Summer 1996): 97–119.

———. *Design and Crime (and Other Diatribes)*. London/New York: Verso, 2002.

Foucault, Michel. *The Birth of the Clinic*. Translated by Alan Sheridan. London: Tavistock, 1973.

———. *Discipline and Punish: The Birth of the Prison*. Translated by Alan Sheridan. London: Allen Lane, 1977.

———. *Power/Knowledge: Selected Interviews and Other Writings*. Edited by Colin Gordon. Translated by Colin Gordon, John Mepham, Kate Soper, and Leo Marshall. New York: Pantheon, 1980, 1981.

———. *Politics, Philosophy, Culture: Interviews and Other Writings 1977–1984*. Edited by Lawrence D. Kritzman. Translated by Alan Sheridan. New York: Routledge, 1988.

———. *Language, Counter-memory, Practice: Selected Essays andInterviews*. Edited by Donald F. Bouchard. Ithaca, New York: Cornell University Press, 1996.

———. *Ethics: Subjectivity and Truth*. Edited by Paul Rabinow. Translated by Robert Hurley et al., London: Allen Lane, 1997.

———. *Aesthetics, Method, and Epistemology*. Edited by James D. Faubion. Translated by Robert Hurley et al. In *Essential Works of Foucault, 1954–1984*, volume 2. New York: The New Press, 1998.

———. *Foucault Live: Interviews, 1966–1984*. Edited by Sylvère Lotringer. Translated by J. Johnston. New York: Semiotext(e), 1989.

———. *The History of Sexuality*. Translated by Robert Hurley. Harmondsworth, Middlesex, England: Penguin, 1990 rpt.

———. *The Archaeology of Knowledge*. Translated by Alan Sheridan. London: Tavistock Publications, 1994.

———. *Security, Territory, Population: Lectures at the Collège de France, 1977-1978*. Edited by Michel Senellart. Translated by Graham Burchell. Hampshire/New York: Palgrave Macmillan, 2007.

Frascari, Marco. *Monsters of Architecture: Anthropomorphism in Architectural Theory*. Lanham, MD: Rowman & Littlefield Publishers, 1991.

Frazer, John. *An Evolutionary Architecture*. London: Architectural Association, 1995.

Freyer, Margaret. "The Bombing of Dresden." In John Carey (ed.), *Eyewitness to History*, 608–11. New York: Avon Books, 1987.

Friedman, Yona. *Programmes and Manifestoes on Twentieth-Century Architecture*. Edited by Ulrich Conrads. London: Lund Humphries, 1970.

Fry, Tony. "Design, Ethics and Identity." *Design Philosophy Papers* 3 (2006): 1–3.

Gans, Deborah, and Zehar Kuz. *The Organic Approach to Architecture*. New York/Chicester: Wiley, 2003.

Gasché, Rodolphe. *Inventions of Difference: On Jacques Derrida*. Cambridge, MA: Harvard University Press, 1995.

Gatens, Moira. *Imaginary Bodies: Ethics, Powers and Corporeality*. London/New York: Routledge, 1996.

Gibson, William. *Neuromancer*. New York: Ace, 1984.

Ginsburg, Faye D. "Screen Memories: Resignifying the Traditional Indigenous Media." In Faye D. Ginsburg, Lila Abu Lughod, and Brian Larkin (eds.), *Media Worlds: Anthropology on a New Terrain*, 39–58. Berkeley: University of California Press, 2002.

Goldschmidt, Richard. *The Material Basis of Evolution*. New Haven, CT: Yale University Press, 1940.

Grayling, Anthony, Clifford. *Among the Dead Cities: Was the Allied Bombing of Civilians in WWII a Necessity or a Crime?* London: Bloomsbury, 2006.

Greenhalgh, Paul. *Ephemeral Vistas: The Expositions Universelles, Great Exhibitions and World Fairs, 1851–1939*. Manchester: Manchester University Press, 1988.

Grierson, Elizabeth. "The Politics of Knowledge: A Poststructuralist Approach to Visual Arts Education in Tertiary Sites." PhD dissertation, The University of Auckland, New Zealand, 2000.

———. "Utopia for All Seasons: Futuring Art—Futuring Education." In Irene Barberis (ed.), *The Chicago Project: Fold-Out Futures II*, 7–14. Melbourne: RMIT Publishing, 2008.

———. "Territories: Contemporary Photographic Work from Australia and China." *Territories*, Project Space, Melbourne. Last modified November 5, 2010. Accessed February 28, 2010. http://schoolofartgalleries.dsc.rmit.edu.au/PSSR/exhibitions/2009/territories/territories.pdf.

———. "Building Dwelling Thinking and Aesthetic Relations in Urban Spaces: A Heideggerian Perspective on Relational Pedagogy as a Form of Disclosure." *ACCESS: Critical Perspective on Communication, Cultural & Policy Studies* 29, no. 1 (2010): 25–41.

———. "Ways of Deconstructing." In Elizabeth Grierson and Laura Brearley (eds.), *Creative Arts Research: Narratives of Methodologies and Practices*, 149–63. Rotterdam: Sense Publishers, 2009.

Grierson, Elizabeth, and Laura Brearley, eds. *Creative Arts Research: Narratives of Methodologies and Practices*. Rotterdam: Sense Publishers, 2009.

Grierson, Elizabeth, and Michael A. Peters. "Introduction: The Legacy of Jacques Derrida." *The Legacy of Jacques Derrida, ACCESS Critical Perspectives on Communication, Cultural & Policy Studies* 24, no. 1 & 2 (2006): 3–14.

Guattari, Félix. *Chaosmosis: An Ethico-Aesthetic Paradigm*. Translated by Paul Bains and Julian Pefanis. Sydney: Power Publications, 1995.

———. *The Three Ecologies*. Translated by Ian Pindar and Paul Sutton. London: The Athlone Press, 2000.

Haber, Honi F. *Beyond Postmodern Politics: Lyotard, Rorty, Foucault*. New York: Routledge, 1994.

Hall, Maria Boas. *The Scientific Renaissance, 1450–1630*. London: Dover, 1994.

Hall, Stuart, ed. *Representations: Cultural Representation and Signifying Practices*. London: Sage and Open University Press, 1997.

Halbwachs, Maurice. *Les Cadres Sociaux de la Memoire*. Paris: Alcan, 1925.

Halloway, Tim. "The Fire-bombing of Dresden: An eyewitness account." *Timewitnesses*. Accessed November 29, 2007. http://timewitnesses.org/english/~lothar.html.

Hamby, Louise. "A Question of Time: *Ten Canoes*." *The Australian Journal of Anthropology* 18 no. 1 (2007): 123–26.

Hansen, Mark. "Internal Resonance, or Three Steps towards a Non-Viral Becoming." *Culture Machine* 3 (2001). http://www.culturemachine.net/index.php/cm/article/viewArticle/429/446.

Haraway, Donna. *When Species Meet*. Minneapolis: University of Minnesota Press, 2008.

Hardt, Michael, and Antonio Negri. *Multitude: War and Democracy in the Age of Empire*. New York: Penguin, 2004.

Haweis, Mary Eliza. *The Art of Beauty*. Boston: Adamant Media, 1878/2005.

Hawker, Philippa. "Canoe Culture Bridges Gaps." *The Age*, June 3, 2006.

Hawley, Charles. "Dresden Bombing Is To Be Regretted Enormously." *Spiegel Online*. Last modified February 1, 2005. http://www.spiegel.de/international/0,1518,341239,00.html.

Hayes, Gary "What Have You Got Against My Bot?" Personalize Media blog. Last modified February 18, 2008. http://www.personalizemedia.com/index.php/2008/02/18/377/.

Heer, Rolf de, and Peter Dijgirr, dirs. *Ten Canoes*. Palace Films. Australia, 2006.

Hegel, G. W. F. *Phenomenology of Spirit*. Translated by Arnold V. Miller. Oxford: Oxford University Press, 1977.

———. *Hegel: Introductory Lectures on Aesthetics*. Translated by Bernard Bosanquet. London: Penguin, 1993.

Heidegger, Martin. *The Question Concerning Technology and Other Essays*. Translated by William Lovitt. New York: Harper & Row, 1977.

———. *An Introduction to Metaphysics*, Translated by Ralph Manheim. New Haven, CT/ London: Yale University Press, 1987/1959.

———. *Nietzsche: Volumes I & II*. Translated by David F. Krell. San Francisco: Harper & Collins, 1991.

———. "The Origin of the Work of Art." In *Basic Writings: From "Being and Time" (1927) to "The Task of Thinking" (1964)*. Edited by David F. Krell, 139–212. London: Routledge, 1993.

———. *The Fundamental Concepts of Metaphysics: World,Finitude, Solitude*. Translated by William McNeill and Nicholas Walker. Bloomington/Indianapolis: Indiana University Press, 1995.

———. *Hölderlin's Hymn* "The Ister." Translated by William McNeill and Julia Davis. Bloomington/Indianapolis: Indiana University Press, 1996.

———. *Being and Time: A Translation of Sein und Zeit*. Translated by Joan Stambaugh. Albany: SUNY, 1996.

———. "Being and Time." In *Basic Writings: From "Being and Time" (1927) to "The Task of Thinking" (1964)*. Edited by David F. Krell, 37–91. London: Routledge, 1999.

———. "Building Dwelling Thinking." In *Basic Writings: From "Being and Time" (1927) to "The Task of Thinking" (1964)*. Edited by David F. Krell. Tranlslated by Frank Capuzzi, 343–64. London: Routledge, 1999.

Hensel, Michael, Achim Menges, and Michael Weinstock, eds. *Emergence: Morphogenetic Design Strategies, AD* 74, no. 3 (May/June, 2004).

———. *Techniques and Technologies in Morphogenetic Design AD* 76, no. 2 (March/April 2006).

Hensel, Michael, and Achim Menges. *Versatility and Vicissitude: Performance in Morpho-Ecological Design, AD* 78, no. 2 (March/April 2008).

Hirsch, Marianne. "Postmemories in Exile." *Poetics Today* 17, no. 4, *Creativity and Exile: European/American Perspectives* II (Winter 1996): 659–86.

———. "Surviving Images: Holocaust Photographs and the Work of Postmemory." *Yale Journal of Criticism* 14, no. 1 (2001): 5–37.

Holland, John. *Emergence: From Chaos to Order*. Oxford: Oxford University Press, 1998.

Holden, Maureen K. "Virtual Environments for Motor Rehabilitation: Review." *CyberPsychology & Behavior* 8, no. 3 (June 1, 2005): 187–211. doi: 10.1089/cp b.2005.8.187.

Howe, Nicholas, "Kilroy in Dresden," *Dissent* (Spring 2001). Last modified March 3, 2011. http://www.dissentmagazine.org/article/?article=1267.

Howells, C. *Derrida: Deconstruction from Phenomenology to Ethics*. Oxford: Polity Press, 1999.

Huyssen, Andreas. *Present Pasts: Urban Palimpsests and the Politics of Memory*. Stanford: Stanford University Press, 2003.

Ingraham, Catherine. "The Faults of Architecture: Troping the Proper." *Assemblage* 7 (1988): 7–13.

———. "Animals 2: the problem of Distinction." *Assemblage* 14 (1991): 25–29.

———. *Architecture, Animal, Human: The Asymmetrical Condition*. London: Routledge, 2006.

Jablonka, Eve, and Marion J. Lamb. *Evolution in Four Dimensions*. Cambridge, MA: MIT Press, 2005.

Jacobus, Mary, Evelyn Fox Keller, and Sally Shuttleworth, eds. *Body/Politics: Women and the Discourses of Science*. London/New York: Routledge, 1990.

Janser, Andres. "Only Film Can Make the New Architecture Intelligible! Hans Richter's Die Neue Wohnung and the Early Documentary Film on Modern Architecture." In François Penz and Maureen Thomas (eds.), *Cinema and Architecture: Melies, Mallet-Stevens, Multimedia*, 34–49. London: British Film Institute, 1997.

Jencks, Charles. *The Language of Post-modern Architecture*. London: Academy Editions, 1978.

Jenkins, Henry. *Convergence Culture: Where Old and New Media Collide*. New York: New York University Press, 2006.

Jenks, Christopher. "The Centrality of the Eye in Western Culture." In Christopher Jenks (ed.), *Visual Culture*. London: Routledge, 1995.

Jennings, Michael W., and H. Eiland, eds. *Walter Benjamin: Selected Writings* 4, 1938–40. Cambridge, MA/London: Belknap Press, 2003.

Johnson, Steven. *Emergence: The Connected Lives of Ants, Brains, Cities and Software*. London: Penguin, 2001.

Kant, Immanuel. *The Critique of Judgement*. Translated by James Creed Meredith. Oxford: Clarendon Press, 1952.

Kasulis, T. P. *Zen Action/Zen Person*. Honolulu: University Press of Hawaii, 1981.

Kay, Lily E. "How a Genetic Code Became an Information System." In Agatha C. Hughes and Thomas P. Hughes (eds.), *Systems, Experts and Computers*, 463–91. Cambridge, MA: MIT Press, 2000.

Keller, Evelyn Fox. *Refiguring Life: Metaphors of Twentieth-Century Biolog y*. New York: Columbia Press, 1995.

Krippendorff, Klaus. *The Semantic Turn: A New Foundation for Design*. Boca Raton, FL: Taylor & Francis, 2006.

Kuhn, Annette, and Kirsten Emiko McAllister, eds. *Locating Memory: Photographic Acts 4*. New York: Berghahn Books, 2006.

Kwinter, Stanford. "Leap in the Void: A New Organon?" In Cynthia Davidson (ed.), *Anyhow*. New York: Any Corporation, 1998.

Lakoff, George, and Mark Turner. *Metaphors We Live By*. Chicago/London: University of Chicago Press, 1980.

———. *More Than Cool Reason*. Chicago/London: University of Chicago Press, 1989.

Lally, Sean, and Jessica Young. *Softspace*. London: Routledge, 2007.

Lauer, B. J. "Mapping Information on CD-ROM." Technical papers of the *1991 ACSM ASPRS Annual Convention*. Baltimore: ACSM-ASPRS 2 (1991): 187–93.

Leach, Neil, ed. *Designing for a Digital World*. London: Wiley, 2003.

Leach, Neil, David Turnbull, and Chris Williams, eds. *Digital Tectonics*. London: Wiley, 2004.

Le Dœuff, Michèle. *The Philosophical Imaginary*. Translated by Colin Gordon. London: The Athlone Press, 1989.

Lefèbvre, Henri. *The Production of Space*. Translated by Donald Nicholson-Smith. Oxford: Blackwell, 1991.

Lemm, Vanessa. *Nietzsche's Animal Philosophy: Culture Politics and the Animality of the Human Being*. New York: Fordham University Press, 2009.

Lenoir, Timothy, and Casey Alt. "Flow, Process, Fold." In Antoine Picon and Alessandra Ponte (eds.), *Architecture and the Sciences: Exchanging Metaphors*, 313–52. New York: Princeton University Press, 2003.

Lessig, Lawrence. *Remix: Making Art and Commerce Thrive in the Hybrid Economy*. New York: Penguin, 2008.

Levinas, Emmanuel. "The Trace of the Other." In Mark C. Taylor (ed.), *Deconstruction in Context*, translated by Alphonso Lingis, 345–59. Chicago: University of Chicago Press, 1986.

———. *Time and the Other: And Additional Essays*. Translated by Richard A. Cohen. Pittsburgh: Duquesne University Press, 1987.

———. *Outside the Subject*. Translated by Michael B. Smith. Stanford: Stanford University Press, 1994.

———. *Ethics and Infinity: Conversations with Philippe Nemo*. Translated by Richard A. Cohen. Pittsburgh: Duquesne University Press, 1995.

———. "Wholly Otherwise." In *Re-Reading Levinas*. Edited by Robert Bernasconi and Simon Critchley, 3–10. Bloomington/Indianapolis: Indiana University Press, 1991.

———. *Totality and Infinity: An Essay on Exteriority*. Translated by Alphonso Lingis. Pittsburgh: Duquesne University Press, 2001/1969.

———. *Otherwise than Being or Beyond Essence*. Translated by Alphonso Lingis, Pittsburgh: Duquesne University Press, 2006.

Lévi-Strauss, Claude. *The Savage Mind*. Chicago: University of Chicago Press, 1968.

Loo, Stephen. "Responding by Mimicry, or Three Frames towards Becoming a Visual Animal." In Ian North (ed.), *Visual Animal: Crossovers, Evolution and New Aesthetics*, 161–73. Adelaide: Contemporary Art Centre of South Australia, 2007.

———. "Emergent Molarities: Resistances on the Molecular Plane of Biology and Digital Architecture." *Architectural Theory Review* 17, no. 1 (2012): 60–75.

Lozano-Hemmer, Raphael. "Interview by José Luis Barrios." *Subsculptures*. Translated by Rebecca MacSween. Last modified September 26, 2005, http://www.lozano-hemmer.com/texts/bibliography/articles_interviews_essays/Subsculptures_2005jlb.pdf.

Lynn, Greg. "Architectural Curvilinearity: The Folded, the Pliant and the Supple." In Greg Lynn (ed.), *Folding in Architecture, AD* 63, no. 3/4 (March/April, 1993): 8–15.

———. *Animate Form*. New York: Princeton Architectural Press, 1999.

———. "Greg Lynn: Embryological Houses." In Ali Rahim (ed.), *Contemporary Processes in Architecture, AD* 70, no. 3 (2000): 26–35.

———. *Folds, Bodies and Blobs: Collected Essays*. Brussells: La Lettre Vole, 2004.

Lynn, Greg, and Ingeborg M. Rocker. "Calculus Based Form: An Interview with Greg Lynn." *Programming Cultures, AD* 76, no. 4 (July/August, 2006).

Lyotard, Jean-François. *The Differend: Phrases in Dispute*. Minneapolis: University of Minnesota Press, 1988.

———. *Political Writings*. Translated by Bill Readings and Kevin P. Geiman. Minneapolis: University of Minnesota Press, 1983.

———. *The Postmodern Condition: A Report on Knowledge*. Translated by Geoff Bennington and Brian Massumi. Minneapolis: University of Minnesota Press, 1984.

———. *The Differend: Phrases in Dispute*. Translated by Georges Van Den Abbeele. Oxford: Manchester University Press, 1988.

Macdonald, Gaynor. "Photos in the Wiradjuri Biscuit Tins: Negotiating and Validating Colonial Histories." *Oceania* 73 (2003): 225–42.

MacLuhan, Marshall. *Understanding Media*. London: Routledge and Kegan Paul, 1974.

Major Projects Victoria. "Our Approach." Last modified August 16, 2010. http://www.majorprojects.vic.gov.au/about/our-approach2.

Malpas, J. E. *Place and Experience*. Cambridge: Cambridge University Press, 1999.

Manovich, Lev. *The Language of New Media*. Cambridge, MA: MIT Press, 2001.

Manzo, L. C. "Beyond House and Haven: Towards a Revisioning of Emotional Relationships with Places." *Journal of Environmental Psychology* 23, no. 1 (March 2003): 47–61.

Margolin, Victor. *The Politics of the Artificial*. Chicago/London: University of Chicago Press, 2002.

Margulis, Lynn, and Dorion Sagan. *Acquiring Genomes: A Theory of the Origins of Species*. New York: Basic Books, 2002.

———. *Microcosmos: Four Billions of Microbial Evolution*. Berkeley: University of Berkeley Press, 1986.

Margulis, Lynn. *Origin of Eukaryotic Cells: Evidence and Research Implications for a Theory of the Origin and Evolution of Microbial, Plant, and Animal Cells on the Precambrian Earth*. New Haven, CT: Yale University Press, 1970.

Marsh, Georgia. "An Interview with Christian Boltanski." *Postmemories in Exile Poetics Today* 17, no. 4, *Creativity and Exile: European/American Perspectives* II (Winter 1996): 659–86.

Martin, Reinhold. "Critical of What?: Toward a Utopian Realism." *Harvard Design Magazine* 22 (Spring/Summer 2005): 2.

Martin, Susan. "'Surmounted by Stuffed Sheep' Exhibitions and Empire in Nineteenth-century Australian Women's Fiction." *Exhibition, Australia and the World Seize the Day* 1, no. 1. Edited by Kate Darian-Smith, Richard Gillespie, Caroline Jordan, and Elizabeth Willis. Clayton: Monash University ePress (September 2008), 121–1211.

Mason, J. *Researching Your Own Practice: The Discipline of Noticing*. London: Routledge Falmer, 2002.

Matchett, Freda. *Krsna, Lord or Avatara? The Relationship between Krsna and Visnu in the Context of the Avatara Myth as Presented by the Harivamsa, the Visnupurana and the Bhagavatapurana*. New York: Routledge, 2000.

McQuire, Scott. *The Media City: Media, Architecture and Urban Space*. London: Sage, 2008.
Menges, Achim, ed. *AD: Material Computation: Higher Integration on Morphogenetic Design*. London: Wiley, March 2012.
Migaryou, Frederic, and Zeynap Mennan. *Architectures Non Standard*. Paris: Editions du Centre Pompidou, 2003.
Milgram, Paul, and Fumio Kishino. "A Taxonomy of Mixed Reality Visual Displays." *IEICE Transactions on Information Systems* E77-D, no. 12 (December 1994).
Mitchell, William. *Me++: The Cyborg Self and the Networked City*. Cambridge, MA: MIT Press, 2003.
Montana, Andrew. *The Art Movement in Australia*. Melbourne: The Miegunyah Press, 2000.
Moretti, Franco. *Atlas of the European Novel*. London/New York: Verso, 1998.
Morris, William. *Prose, Verse, Lectures and Essays*. Edited by G. D. H. Cole. London: Nonesuch Press, 1946.
Mouffe, Chantal, ed. *Dimensions of Radical Democracy: Pluralism, Citizenship, Community*. London: Verso, 1992.
Mouffe, Chantal. *The Return of the Political*. London: Verso, 1993.
Murray, Janet Horowitz. *Hamlet on the Holodeck: The Future of Narrative in Cyberspace*. Cambridge, MA: MIT Press, 1998.
Negroponte, Nicholas. *Being Digital*. New York: Vintage, 1995.
Nietzsche, Friedrich. "On the Uses and Disadvantages of History for Life." In Daniel Breazeale (ed.), *Nietzsche: Untimely Meditations*, 57–123. Cambridge: Cambridge University Press, 1997.
———. *Human, All Too Human*. Translated by Marion Faber and Stephen Lehmann. Nebraska: University of Nebraska Press, 1986.
Norris, Christopher, and Paul de Man. *Deconstruction and the Critique of Aesthetic Ideology*. New York: Routledge, 1988.
Norris, Christopher, and Andrew Benjamin. *What Is Deconstruction?* London: Academy Editions, 1996.
Olssen, Mark. *Michel Foucault: Materialism and Education*. Westport, CT/London Bergin & Garvey, 2006.
Orr, Kirsten. "A Force for Urbanism and National Identity: Nineteenth-Century Australian International Exhibitions and their Domestic Exhibits." *Proceedings of XIXth Conference of the Society of Architectural Historians Australia and New Zealand (SAHANZ)*, Brisbane, October 4–7, 2002.
Peers, Laura, and Alison K. Brown, eds. *Museum and Source Communities: A Routledge Reader*. London: Routledge, 2003.
Peters, K. "The Contemporary Landscape in Art." Paper presented at *AAG 2007 Annual Meeting*, San Francisco, April 17–21, 2007.
Pinney, Christopher. "The Parallel Histories of Anthropology and Photography." In Elizabeth Edwards (ed.), *Anthropology and Photography 1860–1920*. New Haven, CT: Yale University Press, 1992.
———. "Introduction: How the Other Half. . . ." In Christopher Pinney and Nicolas Peterson (eds.), *Photography's Other Histories*. Durham/London: Duke University Press, 2003.
Poignant, Roslyn, and Axel Poignant. *Encounter at Nagalarramba*. Canberra: National Library of Australia, 1996.
Prentice, Stephen. "How to Justify Enterprise Investment in Virtual Worlds." *Gartner Business Analysis*. Last modified March 20, 2008. http://www.gartner.com/DisplayDocument?ref=g_search&id=627708&subref=simplesearch.
Presner, Todd Samuel. "'What a Synoptic and Artificial View Reveals': Extreme History and the Modernism of W. G. Sebald's Realism." *Criticism* 46, no. 3 (Summer 2004): 341–60.
Rendell, Jane. *Art and Architecture: A Place Between*. London/New York: I. B. Tauris, 2006.
Reynolds, Molly, Tania Nehme, and Rolf de Heer. *The Balanda and the Bark Canoes*. Fandango Australia, 2006.
Rigsby, Bruce, and Nicolas Peterson, eds. *Donald Thomson: The Man and the Scholar*. Canberra: The Academy of the Social Sciences in Australia, 2005.
Robinson, Tim. *Stones of Aran: Pilgrimage*. London: Penguin (Non-Classics), 1991.

Rybicki, Joe. "The Real and the Semi-Real." 1Up. Last modified April 12, 2006. http://www.1up.com/do/blogEntry?bId=6883235&publicUserId=4553267.

Rymaszewski, Michael, Wagner James Au, Mark Wallace, Catherine Winters, Cory Ondrejka, and Benjamin Batstone-Cunningham. *Second Life: The Official Guide*. New Jersey: Wiley, 2007.

Rystedt, B. "Compact Disks for Distribution of Maps and Other Geographic Information." *Proceedings of 13th ICC Conference*, Morelia, Mexico, International Cartographic Association (ICA) IV (1987): 479–84.

Salomon, Mandy. "Business in Second Life." Paper presented as Smart Internet Technology CRC Research Paper. Last modified December 22, 2010. Accessed June 18, 2008. http://www.smartservicescrc.com.au/Outcomes.html.

Schofield, Damian. "Animating and Interacting with Graphical Evidence: Bringing Courtrooms to Life with Virtual Reconstructions." Paper presented at *IEEE Conference on Computer Graphics, Imaging and Visualisation*, Bangkok, Thailand, August 14–16 2007.

Sebald, W. G. *Austerlitz*. New York: The Modern Library, 2001

Seibert, Detlef. "British Bombing Strategy in World War Two." BBC History. Last modified February 17, 2011. http://www.bbc.co.uk/history/worldwars/wwtwo/area_bombing_01.shtml.

Sharp, Darren. "Digital Lifestyles Monitor." Paper presented as *Smart Internet Technology CRC* Research Paper. Last modified December 22, 2010. http://www.smartservicescrc.com.au/Outcomes.html.

Shea, Kristina. "Creating Synthesis Partners." *AD* 72, no. 1 (2002): 42–5.

Sherman, William, and Alan Craig. *Understanding Virtual Reality: Interface, Application, and Design*. San Franscicso: Morgan Kaufman Publishers, Elsevier Science, 2003.

Simmel, Georg. *The Metropolis and Mental Life (1903)*. Adapted by D. Weinstein from *The Sociology of Georg Simmel*. Translated by Kurt Wolff. New York: Free Press, 1950, 409–424. http://www.altruists.org/f792.

Sontag, Susan. *On Photography*. New York: Delta, 1977.

———. *Illness as a Metaphor*. New York: Farrar, Straux and Giroux, c. 1978.

———. *Aids and Its Metaphors*. London: Penguin, 1988.

Sparke, Penny. *The Modern Interior*. London: Reaktion Books, 2008.

Speaks, Michael. "It's Out There . . . the Formal Limits of the American Avant-Garde." *Architectural Design: Hypersurface Architecture* 68, 5/6 (1998): 26–31.

———. "No Hope, No Fear." *ARQ* 6/3, 209–12.

Spitzer, Leo. "Back Through the Future: Nostalgic Memory and Critical Memory in a Refuge from Nazism." In Mieke Bal, Jonathan Crew, and Leo Spitzer (eds.), *Acts of Memory*, 87–104. Hanover: University Press of New England, 1999.

Soja, Edward. *Postmodern Geographies: The Reassertion of Space in Social Theory*. London: Verso, 1989.

———. *Thirdspace: Expanding the Geographical Imagination*. Oxford: Blackwell, 1996.

Spuybroek, Lars. *NOX: Machining Architecture*. London: Thames & Hudson, 2004.

Stewart, Susan. "Prologue: From the Museum of Touch." In Marius Kwint, Christopher Breward, and Jeremy Aynesley (eds.), *Material Memories, Design and Evocation*. London: Berg, 1999.

Sutton, Peter. "Science and Sensibility on a Foul Frontier: At Flinders Island, 1935." In Bruce Rigsby and Nicolas Peterson (eds.), *Donald Thomson: The Man and the Scholar*. Canberra: The Australian Academy of the Social Sciences in Australia, 2005.

Sveistrup, Heidi. "Motor Rehabilitation Using Virtual Reality." *Journal of NeuroEngineering and Rehabilitation* 1, no. 10 (2004). doi: 10.1186/1743-0003-1-10.

Taminiaux, Jacques. "The Origin of 'The Origin of the Work of Art.'" In *Poetics, Speculation and Judgement: The Shadow of the Work of Art from Kant to Phenomenology*. Translated by Michael Gendre, 153–70. Albany: SUNY Press, 1993.

Tate, Audrey. *Ada Cambridge: Her Life and Work 1844–1926*. Melbourne: Melbourne University Press, 1991.

Taylor, Frederick. *Dresden: Tuesday, February 13, 1945*. New York: Harper Collins, 2004.

Taylor, William. "Building on Transience: Tolerance and the Subjective Dimension of Technology." Paper presented at *Techniques and Technologies, Transfers and Transformation, Association of Australasian Schools of Architecture (AASA) International Education Conference*. Sydney: University of Technology Sydney, 2007.

Ten Canoes "Press Kit." Palace Films Australia, 2006. Last modified July 1, 2010. http://www.tencanoes.com.au/tencanoes/info.htm.

Thacker, Eugene. *The Global Genome*. Cambridge, MA: MIT Press, 2006.

Thompson, D'Arcy Wentworth. *On Growth and Form*. New York: Dover Publications, 1992.

Thomson, Donald. *Donald Thomson in Arnhem Land*. Melbourne: The Miegunyah Press, 2006.

Tuana, Nancy. "The Weaker Seed: The Sexist Bias of Reproductive Theory." In Nancy Tuana (ed.), *Feminism and Science*, 147–71. Bloomington/ Indianapolis: Indiana University Press, 1989.

Vaughan, Laurene. "Emplacing Local Invention." *Studies in Material Thinking Journal* 1, no. 2 (April, 2008). Last modified June 25, 2008. http://www.aut.ac.nz/material_thinking/materialthinking2/currentissue.html.

Vidler, Anthony. "From Anything to Biothing." In Cynthia C. Davidson (ed.), *Anything*, 226–32. Cambridge, MA: MIT Press, 2001.

Virilio, Paul. *The Information Bomb*. Translated by Chris Turner. New York: Verso, 2005.

Von Hippel, Eric. *Democratizing Innovation*. Cambridge, MA: MIT Press, 2005.

Wall, E. "Walking: Tim Robinson's Stones of Aran." *New Hibernia Review* 12, no. 3 (Fómhar/ Autumn 2008): 66–79.

Wall, Thomas Carl. *Radical Passivity: Levinas, Blanchot, and Agamben*. New York: SUNY Press, 1999.

Wardrip-Fruin, Noah, and Pat Harrigan, eds. *First Person: New Media as Story, Performance and Game*. Cambridge, MA: MIT Press, 2004.

Warner, Marina, ed. *Eyes, Lies and Illusions*. Melbourne: Australian Centre for the Moving Image (ACMI), 2006.

Weinstock, Michael. *The Architecture of Emergence: The Evolution of Form in Nature and Civilisation*. London: Wiley, 2010.

Wertheim, Margaret. *The Pearly Gates of Cyberspace*. New York: W. W. Norton and Co., 1999.

Wigley, Mark. *The Architecture of Deconstruction*. Cambridge, MA/London: MIT Press, 1993.

Willis, Anne M. "Ontological Designing." *Design Philosophy Papers* 2 (2006): 1–11.

Wolfram, Stephen. *A New Kind of Science*. London: Wolfram Media, 2002.

Yeatman, Anna. *Postmodern Revisionings of the Political*. New York: Routledge, 1994.

Young, I. M. *Justice and the Politics of Difference*. Princeton, NJ: Princeton University Press, 1990.

Zaera, Alejandro, and Farshid Moussavi (Foreign Office Architects). *Phylogenesis: FOA's Ark*. Barcelona: Actar, 2003.

Zellner, Peter. *Hybrid Space: New Forms in Digital Architecture*. London: Thames & Hudson, 1999.

Zournazi, Mary. "A Cosmo-Politics–Risk, Hope, Change: A Conversation with Isabelle Stengers." In *Hope: New Philosophies for Change*, 244–74. Annandale: Pluto Press Australia, 2002.

Index

About the Contributors and Editors

CONTRIBUTORS

Yoko Akama is senior lecturer in communication design in the School of Media and Communication, RMIT University. She undertakes research to explore the role and agency of design to tackle social issues. Her expertise is in human-centred design that sees design as "scaffolds" to facilitate communication, engagement, and co-creation with people leading to transformative change. Her current research with the Bushfire CRC conducted a project to strengthen networks within communities to enable more agile and resilient management of bushfires. Through facilitated design interventions, it explored how to initiate engagement, prompt thinking and discussion, build awareness, and reveal tacit knowledge among the community on bushfire planning. She also leads the Service Design Network Melbourne and Design for Social Innovation and Sustainability (DESIS) lab in Melbourne.

Marsha Berry is senior lecturer in the School of Media and Communication at RMIT University, where she lectures in digital media and creative writing for undergraduate degrees. Berry supervises postgraduate research students across a range of topics concerned with new media arts, narrative, design, and mobility. She has numerous publications in digital media and has won international, competitive research grants. Her art practice includes performing arts, poetry, video art, and new media. Recently she has explored notions of memory, place, and displacement through video art, photography, and poetry. Berry's current research investigates social media, perceptions of place, Vietnam, and memory studies.

Karen Burns lectures in history and theory in the architecture program at the University of Melbourne. Her architectural theory and history essays have been published in the journals *Assemblage*, *AD*, *Transition* magazine, and *Architectural Theory Review*, and her writings have been included in the following collections: *Desiring Practices*, *Post Colonial Spaces*, *Intimus*, and *Collectives*. In 2009 she won the prize for best conference paper at the Society of Architectural Historians of Australia and New Zealand (SAHANZ) annual meeting.

William Cartwright, AM, is professor of cartography in the School of Mathematical and Geospatial Sciences at RMIT University, Australia. He joined RMIT after a number of years in the government and private sectors of the mapping industry. He is president of the International Cartographic Association, a Fellow of the Royal Geographical Society, a Fellow of the British Cartographic Society, an Honorary Fellow of the Mapping Sciences Institute Australia and an Honorary Fellow of the Surveying and Spatial Sciences Institute. He holds a Doctor of Philosophy from the University of Melbourne, a Doctor of Education from RMIT University, and six other university qualifications in the fields of cartography, applied science, education, media studies, information and communication technology, and graphic design. He is the author of over 300 academic papers. His major research interest is the application of integrated media to cartography and the exploration of different metaphorical approaches to the depiction of geographical information.

Linda Daley is a senior lecturer in literary and communication studies in the School of Media and Communication at RMIT University, Melbourne. Her research interests span photographic technologies and practices, especially as these relate to Indigenous Australians, literary sociology, and literary philosophy. Her current research focuses on ways in which Indigenous and non-Indigenous encounters are productive of knowledge transfer. She holds a PhD and MA in contemporary European philosophy from the Centre for Comparative Literature and Cultural Studies, Monash University, Melbourne.

Lisa Dethridge writes for film, TV, theater, and various web environments, including *Second Life*. She has taught at New York University; the American Film Institute; University of California, Los Angeles (UCLA); Australian Film, TV and Radio School (AFTRS); and University of Melbourne. She is author of *Writing Your Screenplay* (Allen & Unwin: Sydney, 2003). Dethridge has a PhD in media ecology from New York University, an MA in political science, and a BA (Hons.) in fine arts and literature, University of Melbourne. She is a senior lecturer at RMIT University, Melbourne.

Mark Jackson is associate professor of design for the Faculty of Design and Creative Technologies at AUT University, Auckland, New Zealand. He has held lecturing positions at the University of Adelaide and the University of Sydney. He gained his PhD in architecture in 1994, was a visiting scholar in 1996 at MIT in Boston, visiting professor at the University of Karlsruhe, Germany, in 2003–2004, and visiting scholar at the University of Koblenz-Landaw, Germany, in 2011. Jackson has published in the fields of design history and theory, visual arts, film, media, architecture, and landscape architecture. His current research focus is on ethics and design cultures.

Neil Leach is an architect and theorist. He is currently professor at the University of Southern California. He has also taught at SCI-Arc, Architectural Association, Cornell University, Columbia GSAPP, Dessau Institute of Architecture, University of Bath, and University of Nottingham. He is the author and editor of over twenty books, including *Digital Cities* (Wiley 2009), *Camouflage* (MIT Press 2006), *The Anaesthetics of Architecture* (MIT Press 1999), and *Rethinking Architecture* (Routledge 1997). Leach has been a co-curator of exhibitions at the Beijing Biennial in 2004, 2006, 2008, and 2010, and in Shanghai, 2010, 2011, 2012, and 2013. He is visiting professor at Harvard University, Tongji University, and NASA.

Stephen Loo is professor of architecture at the School of Architecture & Design, University of Tasmania. He has published widely on the spatiality of language, affect, and the biophilosophy of the contemporary subject, which includes ethico-aesthetic models for human action, posthumanist ethics, and experimental digital thinking. Recent publications include *Deleuze and Architecture* (with H. Richot, eds., Edinburgh University Press, 2013); "The (Not So) Smooth Flow between Architecture and Life," in Andrew Ballantyne and Chris L Smith (eds.), *Architecture and the Space of Flows* (Routledge 2012); and "De-signing Ethics: The Good, the Bad and the Performative," in Oksana Zelenko (ed.), *Design and Ethics: Reflections in Practice* (Routledge 2012). His current research project (with Dr. Undine Sellbach) concerns the connections between ethics, psychoanalysis, and the space of the entomological imagination with publications in Angelaki and Parallax.

Scott McQuire has a long-standing interest in interdisciplinary research linking the fields of new media, art, urbanism, and social theory. He is the co-founder of the Spatial Aesthetics research program in the School of Culture and Communication at the University of Melbourne, a member of the executive committee of the Institute for a Broadband-Enabled Society, and one of the initiators of the Research Unit in Public Culture. McQuire is author or editor of seven books and over 100 essays in journals, edited books, and exhibition catalogs. His sole-authored book, *The Media City: Media, Architecture and Urban Space* (Sage 2008) won the 2009 Jane Jacobs Publication Award presented by the Urban Communication Foundation.

Maria O'Connor is senior lecturer in the School of Art and Design, AUT University, Auckland, New Zealand, where she heads the PhD and MPhil programs and coordinates spatial design theoretical studies in undergraduate programs. Her supervisions engage primarily in research projects of a temporal-base spanning art and design installation, performance, and media arts with critical-philosophical underpinnings. Her research foci engage questions of ethics, poetics, and politics in relation to scenes of writing. O'Connor's inscriptive practice questions the limits to the proper of writing across many textual forms and disciplines (film, literature, philosophy, performance, psychoanalysis, criticism, architecture) as an interventional opening to the effects these practices have on the social, political, and ethical subject.

Chris L. Smith is associate dean (Education) and associate professor in Architectural Design and Techné at the University of Sydney. Smith's research is concerned with the interdisciplinary nexus of philosophy, biology, and architectural theory. He has published on the political philosophy of Gilles Deleuze and Félix Guattari (also the subject of his doctoral thesis), technologies of the body, and the influence of the eclipse of Darwinism on contemporary architectural theory. Presently Smith is concentrating upon the changing relation the discourses of philosophy, biology, and architecture maintain with respect to notions of matter and materiality and the medicalization of architecture.

Laurene Vaughan is professor of design and communication in the School of Media and Communication, RMIT University, and a research leader in RMIT Design Research Institute. She was the Nierenmberg Chair, Distinguished Visiting Professor of Design, School of Design, Carnegie Mellon University, 2012–2013. She was project leader and researcher within ACID the Australasian CRC for Interaction Design, 2005–2010. Coming from an art and design education background with a major in sculpture, Vaughan has melded a career of practicing artist, designer, and educator in Australia and internationally. Within her practice she endeavors to explore and present comment on the interactive and situated nature of human experience, particularly creative practice. She has collaborated on major projects and publications exploring the nature of place and design, including *The Stony Rises Project Designing Place* (Melbourne Books) and *Design Collectives: An Approach to Practice* (Cambridge Scholars Press). She is an active member of the Arts and Cartography Commission within the International Cartographic Association.

EDITORS

Elizabeth Grierson is professor of art and philosophy at RMIT University, Melbourne, and was head of the School of Art (2005–2012). Prior to this she held academic positions at AUT University and the University of Auckland, NZ, and was a visiting research fellow at the University of Brighton, UK. She holds a PhD in the philosophy of education and a MA (First Class Hons.) in art history (both University of Auckland), and a Juris Doctor (RMIT). Grierson is a Life Fellow of the Royal Society of Arts (UK), served as World Councillor of International Society of Education through Art (2006–2011), and is editor of *ACCESS* journal (incorporated with *Educational Philosophy and Theory*). She researches and publishes in the philosophy of education, art, and aesthetics, law, and justice. Edited books include *Re-Imagining the City* (Intellect 2013); *Supervising Practices for Postgraduate Research in Art, Architecture and Design* (Sense 2012); *The Doctoral Journey in Art Education* (ASP 2010); *Thinking through Practice* (RMIT 2007, 2008); and *The Arts in Education* (Dunmore 2003); co-authored books include *Designing Sound for Health and Wellbeing* (ASP 2012); *A Skilled Hand and Cultivated Mind* (RMIT 2012, 2008); *Creative Arts Research* (Sense 2009).

Harriet Edquist is professor of architectural history in the School of Architecture and Design at RMIT University, Australia, and was head of the School of Architecture (2001–2007). She has published numerous books on Australian architecture, art, and design, most recently *Michael O'Connell. The Lost Modernist* (2011) and *Pioneers of Modernism. The Arts and Crafts Movement in Australia* (2008), and co-authored *A Skilled Hand and Cultivated Mind* (RMIT 2012, 2008). She is foundation director of the RMIT Design Archives, a facility dedicated to the collection and dissemination of archives representing historical and contemporary practices, which together tell the story of Melbourne as a design city. The Archives supports the work of scholars by providing the resources and support for research into Melbourne's designed environment, design professions, and design practices.

Hélène Frichot is assistant professor in critical studies in architecture, School of Architecture and Built Environment, KTH, Stockholm. Her first discipline is architecture; she also holds a PhD in philosophy from University of Sydney. Frichot is co-curator (with Esther Anatolitis) of Architecture+Philosophy, a public lecture series and forum (from 2005, Melbourne). Recent publications include "Drawing, Thinking, Doing: From Diagram Work to the Superfold," in *ACCESS* 30:1 (2011); "What Can We Learn from the Bubble Man and his Atmospheric Ecologies," in *IDEA: Interior Ecologies* (2011); "Following Hélène Cixous's Steps Towards a Writing Architecture," in Naomi Stead and Lee Stickells (guest eds.), *ATR (Architecture*

Theory Review), 15:3 (2010); and "On Finding Oneself Spinozist: Refuge, Beatitude and the Any-Space-Whatever," in Charles J. Stivale, Eugene W. Holland, and Daniel W. Smith (eds.), *Gilles Deleuze: Image and Text* (Continuum, 2009). She is editor of *Deleuze and Architecture* (Edinburgh University Press).